Telecom & Networking Glossary

Understanding Communications Technology

by

Aegis Publishing Group, Ltd.

edited by

Robert Mastin

Aegis Publishing Group, Ltd.
796 Aquidneck Avenue
Newport, Rhode Island 02842
www.aegisbooks.com
401-849-4200

Library of Congress Catalog Card Number: 98-45440

International Standard Book Number: 1-890154-09-1
Printed in the United States of America.

10 9 8 7 6 5 4 3

Library of Congress Cataloging-In-Publication Data:

Telecom & networking glossary : understanding communications technology / by Aegis Publishing Group ; edited by Robert Mastin.

p. cm.

ISBN 1-890154-09-1 (pbk.)

1. Telecommunication--Dictionaries. I. Mastin, Robert. II. Aegis Publishing Group Ltd. III. Title: Telecom and networking glossary.

TK5102.T42 1998

621.382 ' 03--dc21 98-45440

CIP

What they're saying about
Telecom & Networking Glossary:

"Voice, video, data and multimedia will be travelling over the same networks—here is the language to understand our information futures."
—*Today's Books*

". . . geared toward the nontechnical person who needs to make sense out of the rapidly growing telecommunications industry." —*Wireless Business & Technology*

"It is difficult to stay abreast of each new telecom acronym that crops up. However, for those telcos making deals and trying to remain competitive, a new tool has arrived to aid your efforts. . . focuses on voice telecom and data networking terminology to aid the nontechnical and small entrepreneur. . . designed to aid the reader in the basic terminology."
—*Telecom Business*

"This easy-to-understand book provides an overview of telecom and networking technology and shows you how it all fits together. . . you'll discover how to evaluate competing technologies to determine which is best for your organization." —*Communication Briefings*

Other Telecom Titles from Aegis Publishing Group:

Telecom Business Opportunities
The Entrepreneur's Guide to Making Money in the Telecommunications Revolution, by Steve Rosenbush
$24.95 1-890154-04-0

Winning Communications Strategies
How Small Businesses Master Cutting-Edge Technology to Stay Competitive, Provide Better Service and Make More Money, by Jeffrey Kagan
$14.95 0-9632790-8-4

Telecom Made Easy
Money-Saving, Profit-Building Solutions for Home Businesses, Telecommuters and Small Organizations, by June Langhoff
$19.95 0-9632790-7-6

The Telecommuter's Advisor
Real World Solutions for Remote Workers, by June Langhoff
$14.95 0-9632790-5-X

900 Know-How
How to Succeed With Your Own 900 Number Business,
by Robert Mastin
$19.95 0-9632790-3-3

The Business Traveler's Survival Guide
How to Get Work Done While on the Road, by June Langhoff
$9.95 1-890154-03-2

Getting the Most From Your Yellow Pages Advertising
Maximum Profits at Minimum Cost, by Barry Maher
$19.95 1-890154-05-9

Digital Convergence
How the Merging of Computers, Communications, and Multimedia Is Transforming Our Lives, by Andy Covell
$14.95 1-890154-16-4

How to Buy the Best Phone System
Getting Maximum Value Without Spending a Fortune, by Sondra Liburd Jordan
$9.95 1-890154-06-7

1-800-Courtesy
Connecting With a Winning Telephone Image, by Terry Wildemann
$9.95 1-890154-07-5

The Telecommunication Relay Service (TRS) Handbook
Empowering the Hearing and Speech Impaired, by Franklin H. Silverman, Ph.D.
$9.95 1-890154-08-3

Strategic Marketing in Telecommunications
How to Win Customers, Eliminate Churn, and Increase Profits in the Telecom Marketplace,
by Maureen Rhemann
$39.95 1-890154-17-2

The Cell Phone Handbook
Everything You Wanted to Know About Wireless Telephony (But Didn't Know Who or What to Ask), by Penelope Stetz
$14.95 1-890154-12-1

Phone Company Services
Working Smarter with the Right Telecom Tools, by June Langhoff
$9.95 1-890154-01-6

Acknowledgements

This glossary was a collaborative effort with the help of many people. Although it would be impossible to mention everyone who contributed materials, advice, or assistance to this work, I would like to express special appreciation to Candee Wilde, who did the lion's share of the work defining the most important terms in this book.

My heartfelt gratitude goes out to all the nameless government workers who put together the *Glossary of Telecommunications Terms* (Federal Standard 1037C) issued by the General Services Administration. That publication served as the initial basis for many of the definitions here.

Special thanks to June Langhoff, who wrote many of the definitions relating to telephone company services. Lastly, thanks to Dee Lanoue for copy editing and helping make the definitions comprehensible to nontechnical readers.

Robert Mastin
Editor

How to Use This Glossary

Numbers are listed alphabetically based on their spelling. To find "800 number," for example, look under "eight hundred number."

When referring to another term that is defined elsewhere in the glossary, that term will be in **bold** type (only the first time it appears in a given definition) if referring to the bolded term will aid in the comprehension of the current term being consulted. The term is always bold when it follows the words *see* or *see also*.

Most telecom and networking acronyms are listed here. In general, unless the acronym itself is more universally understood than the full term from which it derives, the acronym will refer you to the full term for its definition. For example, "IP" refers you to "Internet protocol," where you will find the description.

Introduction

This glossary is not written for technicians and engineers, nor is it intended to be a comprehensive compilation of *all* telecom and networking terms. There are other glossaries and dictionaries that do a much better job of defining and explaining the more technical terms (see the listing at the end of this introduction). We tried to limit the scope here to offer a big picture of the major enabling technologies, without getting too much into the specialized details that only engineers can love. For example, we will tell you what a *data packet* is and how it fits into the big picture, but we will not tell you how it is built. We have also tried to emphasize the business-related terms and acronyms in common use within this industry.

This glossary is for nontechnical people who need a general understanding of telecommunications and data networking terms. It is written for anyone new to the industry, including new employees, managers, suppliers, vendors, policymakers, decision makers, investors or anyone else who needs a basic understanding of the industry jargon. It is particularly helpful for busy managers of small businesses and organizations that do not have management information system (MIS) or information technology (IT) managers to help make technology decisions. There is nothing more important to any organization than its communications system—its link to the outside world of customers, prospects, suppliers and business partners. If you are in the market to buy, this glossary will help you make the best purchase decision for your organization.

Although every industry has its distinct language, the telecommunications industry is in a class by itself. First, it is huge—a $750 billion global industry—and it is growing fast, as the entire world is in the process of deregulating markets and upgrading infrastructure. The industry is in a constant state of technological advancement and upheaval. Data communications—spawned by the explosion of the Internet and the proliferation of personal computers—are surpassing traditional voice traffic over many networks. Scores of satellites are being launched to provide global wireless coverage. Powerful new technologies are introduced almost daily by the most cutting-edge entrepreneurial companies in the world. New acronyms are coined daily, and it is hard to understand the industry trade press without a program—until this glossary.

This glossary concentrates on explaining the more important terms you will see every day in the trade press or in vendor literature, such as *packet-switching, SONET, DSL,* and *ATM.* It describes not only what the term means, but how it fits into the overall scheme of things and how it compares with other related technologies. The objective here is to describe the higher-level terms as clearly as possible and leave the nitty-gritty technical stuff to other glossaries and dictionaries. You don't need clutter—you need clarity.

We have tried to stay away from computer and electronic terms unless absolutely necessary. While we do in fact include some fairly arcane terms here, the more narrow or specialized the term the briefer the treatment. We are not trying to turn you into an engineer. We simply want you to be able to make sense of the jargon you will encounter as you navigate the often mystifying world of telecommunications.

One of our major objectives was to keep this book short and simple so that it would be inexpensive, widely distributed, and reasonably portable. It is slim enough to fit in your briefcase for ready reference wherever you happen to be.

We are not infallible. Let us know if we made a mistake or left out a term that belongs here. If you can do better and you are particularly ambitious, submit your own proposed definition for any term— we will give you credit if we use it in a later edition. We reprint regularly, and this glossary will evolve along with advances in the technology and changes in the industry. We want to hear from you. E-mail us at *aegis@aegisbooks.com* or write to us at:

Aegis Publishing Group, Ltd.
796 Aquidneck Avenue
Newport, RI 02842

For those of you who need more than is provided in this glossary, try the following resources:

Newton's Telecom Dictionary
by Harry Newton
$29.95, 800 pages
published by Miller Freeman, Inc.
12 West 21st Street, New York, NY 10010
212-691-8215
800-LIBRARY

Encyclopedia of Networking
by Tom Sheldon
$49.99, 1164 pages
published by Osborne McGraw-Hill
2600 Tenth Street, Berkeley, CA 94710

Desktop Encyclopedia of Telecommunications
by Nathan J. Muller
$49.95, 602 pages
published by McGraw-Hill
800-722-4766

Glossary of Telecommunications Terms (Federal Standard 1037C)
published by the General Services Administration
available from Government Institutes, Inc.
4 Research Place, Rockville, MD 20850
301-921-2355

A

access charge A charge made by a local exchange carrier (LEC) for use of its local exchange facilities to originate or terminate traffic that is carried to or from a distant exchange by an interexchange carrier (IXC). Although some access charges are billed directly to local end users, a very large part of all access charges is paid by interexchange carriers. Access charges are often the largest cost component of a given long-distance call.

access code The preliminary digits that a user must dial to be connected to a particular outgoing trunk group or line.

access line A transmission path between user terminal equipment and a switching center.

access node In switching systems, the point where user traffic enters and exits a communications network. Access node operations may include various operations, such as protocol conversion and code conversion.

accounting rate The average call termination charge between two countries. For example, if country "A" charges $3 to terminate calls and country "B" charges $1, the accounting rate is the average of the two, or $2. See also **settlement rate**.

ACD See **automatic call distributor**.

acknowledge character (ACK) A data transmission control character transmitted by the receiving station as an affirmative response to the sending station. An acknowledge character may also be used as an accuracy control character.

acoustic coupler **1.** An interface device for coupling electrical signals by acoustical means—usually into and out of a telephone instrument. The most common instrument features two rubber cups that fit the transmitting and receiving parts of a standard telephone handset. **2.** A terminal device used to link data terminals and radio sets with the telephone network. Such a link is achieved through acoustic (sound) signals rather than through direct electrical connection.

ACU See **automatic calling unit**.

adaptive routing Routing that is automatically adjusted to compensate for network changes such as traffic patterns, channel availability, or equipment failures.

ADCCP See **Advanced Data Communication Control Procedures.**

add-on conference A service feature that allows an additional party to be added to an established call without attendant assistance. A common implementation provides a progressive method that allows a call originator or a call receiver to add at least one additional party.

address resolution protocol (ARP) A Transmission Control Protocol/Internet Protocol (TCP/IP) protocol that dynamically binds a Network-Layer IP address to a Data-Link-Layer physical hardware address, *e.g.*, Ethernet address.

Advanced Data Communication Control Procedures (ADCCP) A bit-oriented Data-Link-Layer protocol used to provide point-to-point and point-to-multipoint transmission of data frames that contain error-control information. ADCCP closely resembles high-level data link control (HDLC) and synchronous data link control (SDLC).

advanced intelligent network (AIN) A broad definition adopted in the early 1990s by most local and long-distance telephone companies to describe the continuing development of a network for the future. The term AIN initially was developed by Bell Communications Research (Bellcore) to define a new type of telephone network that separates the computer programming that controls new telephone services from the programming that controls the switching equipment embedded in the network. The AIN concept has been adopted by all of the regional Bell operating companies and the major long distance providers because it allows them to develop and implement new services for customers much more quickly.

Prior to AIN, the computer programming that controlled the processing and direction of a call was built in to local switching systems. Therefore, in order to offer a new service, changes had to be made to the switching software throughout all switching systems in the network. This could be a costly and time-consuming procedure, one that significantly delayed the rollout of new telephone services. With AIN, the software controlling new service resides in a centralized Service Control Point (SCP)—essentially a database—that can be accessed by many separate switching systems in the network. As a result, changes to the software in a single SCP allows telecommunications providers to introduce new or altered services much more expediently across a wide geographic area in the network.

Another benefit of separating the computer logic controlling each call from the switches in the network has been the ability of third parties, other than the service providers themselves, to create new equipment and services, once again without modifying the switching system software. Since AIN is standardized in North America, all service providers with an AIN network can use AIN-based products. This means calls can be transferred between networks operated by different carriers. Furthermore, the SCP can reside in the carrier's network or on the customer's premises, allowing the customer to determine the routing of individual calls without affecting information in the switching system that pertains to calls to or from other customers.

Various service providers (phone companies) have interpreted AIN in slightly different ways. In general, however a separate signaling system, Signaling System 7 (SS7), is used to carry information between the Service Control Point (SCP), the Signal Switching Points (SSP) and Signal Transfer Points (STP). The call itself is delivered on the regular network. The SSP is a digital switch that pulls information from the SCP database about how a particular telephone call should be routed. The STP is a switch that carries messages back and forth from the SCPs to the SSPs.

advanced mobile phone service (AMPS) The established analog cellular system in the United States, operating in the 800 MHz range using frequency division multiplexing.

Advanced National Radio Data Service (ARDIS) A wireless data communications service for field technicians and mobile workers. Using handheld devices, short data transmissions can be exchanged—using radio signals and a network of some 1,250 radio base stations— with the host computer at company headquarters to access customer records, inventory levels, diagnostic information, order history or any other information needed by the field worker. Originally developed by IBM and Motorola for IBM's field engineers, it is now owned by Motorola and covers more than 400 metropolitan areas in the United States.

Advanced Research Projects Agency Network (ARPANET) A packet-switching network used by the Department of Defense, linking defense facilities, government research laboratories and universities. ARPANET later evolved into the backbone of the Internet.

AIN See **advanced intelligent network.**

AIOD See **automatic identified outward dialing.**

all trunks busy (ATB) An equipment condition in which all trunks (paths) in a given trunk group are busy.

AM See **amplitude modulation.**

ambient noise level The level of acoustic noise existing at a given location, such as in a room, in a compartment, or outdoors. Ambient noise level is measured with a sound level meter, usually measured in dB above a reference pressure level of 0.00002 Pa, *i.e.,* 20 µPa (micropascals) in SI units. A pascal is a newton per square meter.

American National Standards Institute (ANSI) The U.S. standards organization that establishes procedures for the development and coordination of voluntary American National Standards.

American Standard Code for Information Interchange (ASCII) The standard code used for information interchange among data processing systems, data communications systems, and associated equipment in the United States. The ASCII character set contains 128 coded characters, including letters of the alphabet, numbers, punctuation marks and other symbols. Each ASCII character is a 7-bit coded unique character; 8 bits when a parity check bit is included. The ASCII character set consists of control characters and graphic characters. When considered simply as a set of 128 unique bit patterns, or 256 with a parity bit, disassociated from the character equivalences in national implementations, the ASCII may be considered as an alphabet used in machine languages. The ASCII is the U.S. implementation of International Alphabet No. 5 (IA No. 5) as specified in CCITT Recommendation V.3.

amplifier A device use in analog networks to boost the strength of a signal and extend the range of the transmission. Unfortunately, noise is also amplified along with the transmitted signal, making the quality of long-distance analog transmission poor. Repeaters perform the equivalent function in digital networks, but without magnifying noise, resulting in a cleaner signal.

amplitude modulation (AM) Modulation in which the amplitude of a carrier wave is varied in accordance with some characteristic of the modulating signal, such as an audio frequency.

AMPS See **automatic message processing system** or **advanced mobile phone service**.

analog signal A signal that has a continuous nature rather than a pulsed or discrete nature. Electrical or physical analogies, such as continuously varying voltages, frequencies, or phases, may be used as analog signals. For example, an analog signal may vary in frequency, phase, or amplitude in response to changes in physical phenomena, such as sound, light, heat, position, or pressure.

analog transmission Voice signals, including radio, have traditionally been sent using analog transmissions. "Analog" is a shortened form of "analogous," which refers to the fact that the transmitted signal is sent as an electrical signal over twisted pairs of copper wire so that it fluctuates, in terms of volume and pitch, in the same pattern as the original signal. For example, if an oscilloscope monitors the spoken voice, the pattern of waves displayed on the oscilloscope screen would follow the same pattern as (be "analogous to") the modulation of the waves carried over the telephone or broadcast line. Analog transmission is also used to distinguish telecommunications carried on an analog network from those carried on the newer, easier-to-design digital networks.

ANI See **automatic number identification**.

anonymous call rejection This service works in conjunction with caller ID and allows you to automatically reject all calls from callers who have activated blocking caller ID. Your caller receives a recorded message advising that blocked calls are not accepted. Some residential subscribers like this service. Not too handy for the normal business.

ANSI See **American National Standards Institute.**

answer back A signal sent by receiving equipment to the sending station to indicate that the receiver is ready to accept transmission.

answer signal A supervisory signal returned from the called telephone to the originating switch when the call receiver answers the telephone. The answer signal stops the ringback signal from being returned to the caller. The answer signal is returned by means of a closed loop.

application program interface (API) A formalized set of software language and message formats that define how application programs interact with the operating system, with other programs or with communications systems and networks. Programmers use APIs when writing software that will run on a particular operating system or interact with a specific communications system. For example, programmers writing software to run on Windows 95 will use APIs to access icons, drop-down menus, scroll bars, and other graphical tools.

applications service provider (ASP) A new term for a company, often a network service provider or an ISP, that offers services beyond basic transport, such as data storage, database management, enterprise communications, and other applications that can be delivered over the network.

arbitrage In business and finance, this refers to the near simultaneous purchase and sale of securities or currencies in different markets in order to profit from the price differences. In telecommunications, it usually refers to a carrier buying transport minutes low, at wholesale, and selling a portion of those minutes at a higher price to another carrier.

ARDIS See **Advanced National Radio Data Service**.

area code Also known as *service access code*, the three-digit number that represents a "toll" center in the North American Numbering Plan (NANP), comprised of the United States, Canada and the Caribbean. The first set of numbers dialed for a long-distance call (after the "0" or "1"). With the advent of cellular phones, modems, fax machines and data communications, a huge increase in the total number of telephone lines has necessitated a corresponding increase in the number of area codes, going from a fairly static situation through the 1980s (when the middle number was always a "0" or a "1") to a dynamic environment with numerous new codes needed every year to keep up with demand. For an up-to-date list of area codes, see *www.nanpa.com*. For an area code map of the United States, see *www.aegisbooks.com*.

ARP See **address resolution protocol.**

ARPANET See **Advanced Research Projects Agency Network.**

ASCII See **American Standard Code for Information Interchange.**

asynchronous (or asymmetric) digital subscriber line (ADSL) A digital switching technology that uses regular copper telephone lines to carry high-speed data a relatively short distance between telephone company networks and subscribers' homes or offices (the local loop). Asynchronous DSL is so named because data carried downstream, for example from the Internet or another source point, travels through twisted-pair copper wires (regular telephone lines) at theoretical speeds of up to nearly 8.5 Mbps (millions of bits per second) while information sent upstream (from the customer into the telephone company's network), on the other hand, travels at a much slower rate of about 640 Kbps (thousands of bits per second). In practice, however, speeds are now much slower, as telephone companies begin testing new ADSL services. GTE Corp. for example, began offering ADSL in Marina del Rey, Calif., in Nov. 1997, at downstream data rates of 1.5 Mbps and upstream rates of 384 Kbps, at a monthly cost to customers of $99. The reason for slowing ADSL's maximum speeds in both directions is to reduce certain technological difficulties, primarily in homes with older internal wiring systems and, to some extent, to compensate for installation barriers within the telephone network itself, such as distance from the telephone company's central office. This disparity in the upstream and downstream speeds, with the greater bandwidth allocated to the downstream (from source to customer) direction makes sense. In practice, most people simply query the Internet for information—a transaction that requires sending a relatively small amount of information. Downloading a file, however, or sending a full-motion video in a reasonable amount of time, requires far greater bandwidth than the maximum 28.8 Kbps, 33 Kbps or 56 Kbps modems available today. The downstream speeds of most ADSL services are roughly equivalent to data speeds on the T1 lines frequently leased by major corporations. ADSL is undoubtedly the "flavor" of DSL that will be used most for home and small-business customers. Other forms of DSL service include HDSL (high bit-rate DSL); IDSL (ISDN DSL); RADSL (Rate-Adaptive DSL), SDSL (Single-Line DSL); UDSL, (Unidirectional DSL); and VDSL (Very high-data-rate DSL). The definitions of each of these terms are quite technical and most experts agree that ADSL will become the leading xDSL technology. See also **digital subscriber line (DSL)**.

asynchronous operation A sequence of operations in which operations are executed out of time coincidence with any event. An operation that occurs without a regular or predictable time relationship to a specified event.

asynchronous time-division multiplexing (ATDM) Time-division multiplexing in which asynchronous transmission is used.

asynchronous transfer mode (ATM) This is the global standard for integrated broadband transmissions. ATM is described as "integrated" because ATM is equally suited to voice, data and video transmissions. ATM evolved from other packet-switching technologies, although it differs from its predecessors because ATM lines transmits data packets that are fixed in size at 53 bytes each—48 bytes are for user information and the remaining five act as "directions" as to how the packet should be routed over the network to the intended receiver.

In contrast, frame relay is a transmission technology in which the size of the data packets varies during the course of a transmission. Switching experts realized these variable-sized packets would become difficult to manage at the high speeds necessary to carry real-time voice and video, functions for which ATM is well suited. A new 622 Mbps version of ATM will soon replace the current 155 Mbps version on the market. Some ATM backbone networks belonging to carriers operate at speeds up to 2.5 Gbps.

Analog transmissions are synchronous in nature, meaning network capacity goes unused when a line is open but nothing is being transmitted. ATM, on the other hand, allows a combination copper/fiber network to adjust available capacity so that the line is in virtually continuous use, by a variety of customers. In other words, users don't wait in line to send on an ATM line: customers can send any combination of voice, video or data transmission at any time and it will be delivered as soon as there is a gap between other packets on the network. Information in the "header" of the data packet tells the network to route the information in a particular way. All subsequent information that is part of this same transmission will carry that 5-bit header and hence end up at the same destination, where the separated packets are reunited before final delivery to the receiving party.

Interestingly, ATM is the first global transmission standard used by the computer, communications and entertainment industries. It was adopted by the International Telecommunications Union-Telecommunications Services Sector (ITU-T) as the transmission technology for standardized high-speed global communications. Although ATM was initially developed as a wide-are network (WAN) technology, local-area network (LAN) ATM switches have been developed and the technology is making its way closer to users as PBX manufacturers work to incorporate ATM switching into their products. ATM is also a CCITT term for Broadband ISDN using cell-relay transmission protocols.

asynchronous transmission A method of sending information in which each character in the transmission is preceded by a signal that tells the receiving computer to "start" translating digital "0's" and "1's" into information and is ended with a signal telling the receiving computer to stop translating—that this portion of the transmission is over. With synchronous transmissions, on the other hand, computer communication is based on a timing signal that alerts the receiving computer of the precise time the transmission starts and stops. In general, more sophisticated computer systems use synchronous transmission because it is less expensive—the sender doesn't have to add a beginning and ending character (2 bits) to each 8 bits of the transmission string. Each bit of information sent on a network costs the user money. Also, error checking is easier using synchronous transmission methods because one doesn't have to check each group of characters for errors—if an error occurs the entire transmission is re-sent. *Isochronous* transmission refers to simultaneous voice and video transmissions, for example, that must be sent and received in precise sequence in order to be completed properly.

ATM See **asynchronous transfer mode**.

attendant conference A network-provided service feature that allows an attendant to establish a conference connection of three or more users.

attendant console The component of the telephone system used by the operator, or attendant, to field incoming calls and route them to their intended party within the organization. Older consoles feature buttons and codes, while newer ones feature monitors with a graphical user interface (GUI) that shows the status of calls, extensions, holding queues and the like.

attenuation The decrease in intensity of a signal, beam, or wave as a result of energy absorption and scattering (out of the path to the detector, not including the reduction due to geometric spreading), usually expressed in dB.

audiotext (also audiotex) This term broadly describes various telecommunications equipment and services that enable users to send or receive information by interacting with a voice processing system via a telephone connection, using audio input. Voice mail, interactive 800 or 900 programs, and telephone banking transactions are examples of audiotext applications.

automated attendant A device, connected to a PBX, that performs simple voice-processing functions limited to answering incoming calls and routing them in accordance with the touch-tone menu selections made by the caller.

automatic callback A service feature that permits a user, when encountering a busy condition, to instruct the system to retain the called and calling numbers and to establish the call when there is an available line.

automatic call distributor (ACD) A specialized phone system used for handling a high volume of incoming calls, distributing calls to a specific group of terminals. If there are fewer active calls than terminals, the next call will be routed to the terminal that has been idle the longest. An ACD can be programmed a variety of ways. For example, it will send the call to a designated agent, to a recording giving the caller further instructions, or to a voice response unit (VRU). An ACD is much smarter than a **hunt group** that rolls to the first available line in the same order each time, thereby making the first lines always busy while the last lines are only intermittently busy.

automatic calling unit (ACU) A device that enables equipment, such as computers and card dialers, to originate calls automatically over a telecommunications network.

automatic dialer (autodialer) A device that is pre-programmed with frequently dialed phone numbers so that only one or two digits need to be pressed to automatically dial the programmed number. It can be a separate stand-alone device or incorporated into the telephone set.

Automatic Digital Network (AUTODIN) Formerly, a worldwide data communications network of the Defense Communications System, now replaced by the **Defense Switched Network (DSN).**

automatic identified outward dialing (AIOD) A service feature of some switching or terminal devices that provides the user with an itemized statement of usage on directly dialed calls. AIOD is facilitated by automatic number identification (ANI) equipment to provide automatic message accounting (AMA).

automatic message accounting (AMA) A service feature that automatically records data regarding user-dialed calls.

automatic message processing system (AMPS) Any organized assembly of resources and methods used to collect, process, and distribute messages largely by automatic means.

automatic number identification (ANI) A means of identifying the telephone number of the party originating the telephone call. This is accomplished through the use of analog or digital signals that are transmitted along with the call and equipment that can decipher those signals. ANI is essentially **caller ID** for long-distance calls. The only difference is when the information is delivered: with ANI it is before the first ring, with caller ID it is between the first and second rings.

automatic redial A service feature that allows the user to dial, by depressing a single key or a few keys, the most recent telephone number dialed at that instrument. Automatic redial is often associated with the telephone instrument, but may be provided by a PBX, or by the central office. Also called **last number redial.**

automatic ringdown circuit A circuit providing priority telephone service, typically for key personnel; the circuit is activated when the telephone handset is removed from the cradle causing a ringing signal to be sent to the distant unit(s).

automatic route selection (ARS) Electronic or mechanical selection and routing of outgoing calls without human intervention. This is usually a PBX feature that automatically selects the least expensive outgoing trunk based on the outgoing number that was dialed, often using a look-up table of available exchanges and long-distance carriers and their associated costs. Also called *least cost routing (LCR)*.

automatic switching system **1.** In data communications, a switching system in which all the operations required to execute the three phases of information-transfer transactions are automatically executed in response to signals from a user end-instrument. In an automatic switching system, the information-transfer transaction is performed without human intervention, except for initiation of the access phase and the disengagement phase by a user. **2.** In telephony, a system in which all the operations required to set up, supervise, and release connections required for calls are automatically performed in response to signals from a calling device.

Automatic Voice Network (AUTOVON) Formerly, the principal long-haul, unsecure voice communications network within the Defense Communications System, now replaced by the Defense Switched Network (DSN).

available line In voice, video, or data communications, a circuit between two points that is ready for service, but is in the idle state.

B

backbone The high-traffic-density connectivity portion of any communications network. In telecommunications, a backbone is the major—or biggest—pipe (communications line) in the network. Smaller pipes, or lines, feed into this backbone. The backbone is the part of the network that carries the lion's share of traffic. The backbone can also be thought of as a large river into which scores of smaller streams or tributaries feed. The Internet is a wide-area network comprised of a number of backbones—regional networks that carry long-distance traffic. The regional networks comprising the Internet's backbone are connected by network nodes, which are simply switching points.

backhaul In a communications channel, the act of going past the ultimate destination and then doubling back to reach it. For example, a call from New York to Las Vegas may actually go to Los Angeles first and then get backhauled to Las Vegas.

backward channel **1.** In data transmission, a secondary channel in which the direction of transmission is constrained to be opposite that of the primary, or, the forward (user-information) channel. The direction of transmission in the backward channel is restricted by the control interchange circuit that controls the direction of transmission in the primary channel. **2.** In a data circuit, the channel that passes data in a direction opposite to that of its associated forward channel. The backward channel is usually used for transmission of supervisory, acknowledgement, or error-control signals. The direction of flow of these signals is opposite to that in which user information is being transferred. The backward-channel bandwidth is usually less than that of the primary channel.

backward signal A signal sent from the called to the calling station, *i.e.,* from the original data sink to the original data source. Backward signals are usually sent via a backward channel and may consist of supervisory, acknowledgment, or control signals.

band **1.** In communications, the frequency spectrum between two defined limits. **2.** A set of frequencies authorized for use in a geographical area that is defined for common carriers for purposes of communications system management.

bandwidth (BW) All transmission signals, both analog and digital, have bandwidth. Bandwidth is a measure of the difference between the lowest frequency and the highest frequency of a data, video or voice transmission. The term bandwidth, when expressed as bits per second, also is used to measure the quantity of data that a particular transmission line can carry each second.

bandwidth compression **1.** The reduction of the bandwidth needed to transmit a given amount of data in a given time. **2.** The reduction of the time needed to transmit a given amount of data in a given bandwidth. Bandwidth compression implies a reduction in normal bandwidth of an information-carrying signal without reducing the information content of the signal.

baseband network A network in which information is encoded, multiplexed, and transmitted without modulation of carriers. This is accomplished by transmitting direct current electrical pulses directly on a cable, with the presence or absence of voltage representing binary 1s or 0s. This type of transmission degrades easily over distance and is usually limited to local area networks (LANs). For example, Ethernet is a shared baseband network. Contrast this with broadband transmission, where multiple channels are modulated onto separate carrier frequencies, and distances can be great.

baseband signaling Transmission of a digital or analog signal at its original frequencies; *i.e.*, a signal in its original form, not changed by modulation.

base station **1.** A land station in the land mobile service. **2.** In personal communication service, the common name for all the radio equipment located at one fixed location, which serves one or several cells.

basic exchange telecommunications radio service (BETRS) A commercial service that can extend telephone service to rural areas by replacing the local loop with radio communications. In the BETRS, non-government ultra-high frequency (UHF) and very high frequency (VHF) common carrier and the private radio service frequencies are shared.

basic rate interface (BRI) One of the **integrated services digital network (ISDN)** technologies. BRI-ISDN refers to the ISDN configuration intended for small businesses and home customers. The other important ISDN configurations are **broadband ISDN** (B-ISDN) and **primary rate interface** - ISDN (PRI-ISDN), higher-bandwidth versions used in telephone-company and major corporate networks.

BRI-ISDN divides an analog copper line into three digital channels: two 64 Kbps Bearer (or B) channels and one 16 Kbps Delta (or D) channel. The two B channels carry customers' telephone calls, Internet connections, video-conferencing links and other circuit-switched services. The D channel is a packet-switched, data-only line that is shared by users and the telephone company. Carriers use the D channel to carry the information required to direct a call to its destination. (This type of information can be called *control* or *signaling* information.) ISDN customers can also use the D channel to send and receive data.

The total amount of bandwidth BRI-ISDN service can provide is 144 Kbps—significantly more than the 4-kHz analog-based copper lines can support using current modem technology. Most people access the Internet using 28.8 Kbps modems; BRI-ISDN supports Internet access that is roughly four times faster. The extra bandwidth also improves voice service: there is less interference on a digital line, so conversations sound better. In an analog network, call signaling takes place on the same lines that carry voice traffic. An integral part of ISDN is the separate, packet-switched D channel for signaling (See **Signaling System 7**). It takes about one second for an ISDN-supported call to connect, while an analog call might take 20 seconds to go through. And additional speed is just one benefit. Because BRI-ISDN service splits a telephone line into three channels (in a configuration that is often referred to as *2B + D*)—it can support three simultaneous communications. In contrast, an analog line can carry only one conversation or a single data transmission at a time. BRI-ISDN's two B channels carry customers' circuit-switched communications, such as voice calls and faxes; the D channel carries call set-up information for the carrier but can also support packet-switched data transmissions, such as e-mail messages or credit-card authorizations, for the customer.

BRI-ISDN customers can connect as many as eight separate devices—such as telephones, computers, and fax machines—to the same ISDN line, eliminating the need to have separate lines for each communications device. ISDN service also will support separate telephone numbers for each device. In fact, BRI-ISDN can support as many as 64 different telephone numbers, depending on the ISDN switch installed at the carrier's central office.

basic service **1.** A pure transmission capability over a communication path that is virtually transparent in terms of its interaction with customer-supplied information. **2.** The offering of transmission capacity between two or more points suitable for a user's transmission needs and subject only to the technical parameters of fidelity and distortion criteria, or other conditioning.

basic service element (BSE) **1.** An optional unbundled feature, generally associated with the basic serving arrangement (BSA), that an enhanced-service provider (ESP) may require or find useful in configuring an enhanced service. **2.** A fundamental (basic) communication network service; an optional network capability associated with a BSA. BSEs constitute optional capabilities to which the customer may subscribe or decline to subscribe.

basic serving arrangement (BSA) The fundamental tariffed switching and transmission (and other) services that an operating company must provide to an enhanced service provider (ESP) to connect with its customers through the company network.

batched transmission The transmission of two or more messages from one station to another without intervening responses from the receiving station. Also called *batched communications*.

baud The term refers to the information-carrying capacity or the signaling capacity of a communications line. The term, originally created in 1927 to measure the time it took to send one Morse code dot, is falling out of use because it is not specific enough to apply with accuracy to many of today's networks. A baud denotes a single symbol sent over a communications line, but in the case of asynchronous or packet-switched networks, it does not take into consideration the required stop bits. In general, it is more accurate to discuss communications-line speeds in terms of bits per second.

B channel The CCITT designation for a clear channel, 64 Kbps service capability provided to a subscriber under the Integrated Services Digital Network (ISDN) offering. The B channel, also called the bearer channel, is intended for transport of user information, as opposed to signaling information.

bearer channel See **B channel**.

Bell Operating Company (BOC) Any of the 22 operating companies that were divested from AT&T by court order. See also **Regional Bell Operating Company**.

B-ISDN See **broadband ISDN.**

bits per second (b/s or bps) A unit used to express the number of bits passing a designated point per second; a measure of data transmission volume or throughput. Although often equated with transmission speed, this is not technically correct because all data actually travels at the speed of light. The higher the bits per second, however, the faster a given block of data will reach its destination.

bit-stream transmission In bit-stream data transmission, the bits usually occur at fixed time intervals, start and stop signals are not used, and the bit patterns follow each other in sequence without interruption.

bit string A sequence of bits. In a bit stream, individual bit strings may be separated by data delimiters.

block call Depending on your area, this is a service feature that (1) lets you prevent certain types of outgoing calls, such as toll calls (you can override blocking by dialing a personal code) or (2) allows you to program a list of numbers that will be blocked from ringing on your phone line.

BOC See **Bell Operating Company.**

BRI See **basic rate interface**.

bridge In communications networks, a device that (a) links or routes signals from one ring or bus to another or from one network to another, (b) may extend the distance span and capacity of a single LAN system, (c) performs no modification to packets or messages, (d) operates at the data-link layer of the OSI—Reference Model (Layer 2), (e) reads packets, and (f) passes only those with addresses on the same segment of the network as the originating user. A LAN is physically limited to a certain number of nodes and repeaters. A bridge can overcome this limitation by creating another interconnected sub-LAN.

broadband This term is used in a variety of ways. Often, it simply describes a communications line that has a greater bandwidth than a typical voice line (4 kHz). Broadband is often used interchangeably with wideband, which can have many different meanings depending on context. For example, wideband is often used to contrast with narrowband. The network used to deliver cable television (CATV) signals is a considered a true broadband network because the line itself is coaxial cable, which can support a broad range of frequencies, both audio and visual, by dividing the line's total capacity (bandwidth) into many independent channels, each supporting a certain set of frequencies.

broadband ISDN (B-ISDN) An Integrated Services Digital Network (ISDN) offering broadband capabilities. B-ISDN is a CCITT-proposed service that may (a) include interfaces operating at data rates from 150 to 600 Mbps, (b) use asynchronous transfer mode (ATM) to carry all services over a single, integrated, high-speed packet-switched network, (c) have LAN interconnection capability, (d) provide access to a remote, shared disk server, (e) provide voice/video/data teleconferencing, (f) provide transport for programming services, such as cable TV, (g) provide single-user controlled access to remote video sources, (h) handle voice/video telephone calls, and (i) access shop-at-home and other information services. Techniques used in the B-ISDN include code conversion, information compression, multipoint connections, and multiple-connection calls. Current proposals use a service-independent call structure that allows flexible arrangement and modular control of access and transport edges. The service components of a connection can provide each user with independent control of access features and can serve as the basis of a simplified control structure for multipoint and multiconnection calls. Such a network might be expected to offer a variety of ancillary information processing functions.

broadcast networking Where multiple nodes are attached to a LAN, all nodes listen to traffic originated by all other nodes on the network. Hence, the data is broadcast to all network nodes, which ignore all traffic that does not carry their individual destination addresses. Ethernet is such a shared network. The alternative is a point-to-point link that connects the sender to the receiver and does not share the connection with other devices.

broadcast operation The transmission of signals that may be simultaneously received by multiple stations that usually make no acknowledgement.

brouter A combined **bridge** and **router** that operates without protocol restrictions, routes data using a protocol it supports, and bridges data it cannot route.

browser Any computer software program for reading hypertext computer language. Browsers are usually associated with the Internet and the World Wide Web (WWW). A browser may be able to access information in many formats and through different services including HTTP, FTP, Gopher, and Archie. See also **web browser**.

b/s See **bits per second.**

bundled services Since the Telecommunications Act of 1996 freed local and long-distance companies to compete with one another, a key marketing strategy has been to offer "bundled services." This simply means offering a variety of different packages of services designed to appeal, both in price and content, to a specific target market. Before the Telecom Act was passed, telecommunications providers were not allowed to combine local voice, long-distance voice, and data services over a single line. Service providers today can also offer bundled Internet, voice, and data services. The regional Bell companies, faced with increasing competition from competitive local exchange carriers (CLECs) and long-distance providers moving into the local-access business, are fighting back by providing various bundles of local, long-distance, wireless, and Internet service—even cable television service—on one bill. A key component of bundled services is the ability of the telecommunications service provider to bill customers for all of these services with one document (be it electronic or paper). Many consultants consider bundled services and consolidated billing two of the hottest issues in telecommunications competition today. As one analyst says, "[When carriers can offer the] billing and customer care to support bundled service offerings, they maximize customer loyalty and prevent customers from going elsewhere to get the services they need."

burst switching In a packet-switched network, a switching capability in which each network switch extracts routing instructions from an incoming packet header to establish and maintain the appropriate switch connection for the duration of the packet, following which the connection is automatically released.

burst transmission **1.** Transmission that combines a very high data signaling rate with very short transmission times without interruption. **2.** Operation of a data network in which data transmission is interrupted at intervals. Burst transmission enables communications between data terminal equipment (DTEs) and a data network operating at dissimilar data signaling rates.

bus One or more conductors or optical fibers that serve as a common connection for a group of related devices. It is usually depicted graphically as a straight line with various devices connected to the line at different points.

busy signal In telephony, an audible or visual signal that indicates that no transmission path to the called number is available, or that the called number is occupied or otherwise unavailable.

BW See **bandwidth**.

C

cable An assembly of one or more insulated conductors, or optical fibers, or a combination of both, within an enveloping jacket. A cable is constructed so that the conductors or fibers may be used singly or in groups.

cable modem A device that connects a PC or network computer to the cable TV (CATV) network so that the computer may transmit and receive data over the network. Because existing CATV networks already employ high-bandwidth coaxial cable into the home or office, these modems are much faster than dial-up analog modems used on conventional phone lines, offering speeds from 3 to 10 Mbps. (These are practical speeds on a shared network in any given local area, and the more users on the network at the same time the slower the speed—theoretical speed is much higher on a dedicated link. Contrast this with analog modems, which now top out at 56 Kbps under ideal conditions.)

cable TV (CATV) A television distribution method that receives, amplifies, and then transmits signals from distant stations to users via cable (coaxial or fiber) or microwave links. CATV originated in areas where good reception of direct broadcast TV was not possible. Now CATV also consists of a cable distribution system to large metropolitan areas in competition with direct broadcasting. The abbreviation CATV originally meant *community antenna television.* However, CATV is now usually understood to mean *cable TV.*

call **1.** In communications, any demand to set up a connection. **2.** A unit of traffic measurement. **3.** The actions performed by a call originator. **4.** The operations required to establish, maintain, and release a connection. **5.** To use a connection between two stations.

call accounting system (CAS) A system for compiling information about telephone calls to and from a telephone system. It can be a stand-alone device with its own computer or software running on the telephone system computer. It compiles information such as outgoing numbers dialed, circuits used, extensions used, call duration, time of call, and other information useful for monitoring telephone abuse, billing clients, allocating expenses and other management functions.

call attempt In a telecommunications system, a demand by a user for a connection to another user. In telephone traffic analysis, call attempts are counted during a specific time frame. The call-attempt count includes all completed, overflowed, abandoned, and lost calls.

call back A security procedure for identifying a remote calling terminal, whereby the host system disconnects the caller and then dials the authorized telephone number of the remote terminal to reestablish the connection.

call-center A facility staffed with numerous personnel who handle a high volume of either incoming or outgoing telephone calls, or both. Telemarketing operations are *outbound* call-centers. Customer service, technical support or order-taking operations are *inbound* call-centers. The operators are typically networked with computers and specialized software and databases, which help them serve the callers as necessary, from processing orders to troubleshooting a technical problem. See also **screen pop**.

call collision **1.** The contention that occurs when a terminal and **data circuit-terminating equipment (DCE)** specify the same channel at the same time to transfer a call request and handle an incoming call. When call collision occurs, the DCE proceeds with the call request and cancels the incoming call. **2.** The condition that occurs when a trunk or channel is seized at both ends simultaneously, thereby blocking a call.

call completion rate The ratio of successfully completed calls to the total number of attempted calls. This ratio is typically expressed as either a percentage or a decimal fraction.

called-line identification facility A network-provided service feature in which the network notifies a calling terminal of the address to which the call has been connected.

called-party camp-on A communication system service feature that enables the system to complete an access attempt in spite of issuance of a user blocking signal. Systems that provide this feature monitor the busy user until the user blocking signal ends and then proceed to complete the requested access. This feature permits holding an incoming call until the called party is free.

caller ID A network service feature that permits the recipient of an incoming call to determine, even before answering, the number from which the incoming call is being placed. Caller ID lets you see the phone number (and, for an extra fee, the caller's name) of the incoming call before you answer the phone. Caller ID is usually a combination of two features: calling number delivery and calling name delivery. Calling number delivery displays the 10-digit telephone number of an incoming call. Calling name delivery displays the name (as it is listed in the directory) associated with an incoming call. Both services display the date and time—a convenience especially if you are automatically logging calls. The service costs range between $4 and $7.50 per month, depending on the service provider. Software vendors are developing caller ID applications that do **screen**

pops, by selecting a customer's record from the database as the call comes in and displaying a screen full of data such as order history, terms, etc. This allows you to be fully prepared to handle the caller's needs. See also **automatic number identification (ANI)**.

call-failure signal A signal sent in the backward direction indicating that a call cannot be completed because of a time-out, a fault, or a condition that does not correspond to any other particular signal.

call forwarding A service feature, available in some switching systems (central office or PBX), whereby calls can be re-routed automatically from one line, *i.e.,* station number, to another or to an attendant. The service available from the local phone company comes in three basic versions: 1) user programmed forwarding, allowing users to program and reprogram their number to automatically transfer to other designated numbers; 2) busy forwarding, where calls are automatically forwarded when the called line is busy; and 3) delay forwarding, where the call is forwarded if not answered within a predetermined number of rings.

call hold A service feature in which a user may retain an existing call while accepting or originating another call using the same end instrument.

calling card A credit card issued by telephone companies that allows customers to charge telephone calls made from anywhere to their existing home or office accounts. Usually the identifying number is the customer's telephone number followed by a PIN number.

calling-party camp-on A service feature that enables the system to complete an access attempt in spite of temporary unavailability of system transmission or switching facilities required to establish the requested access. Systems that provide calling-party camp-on complete the requested access as soon as the necessary facilities become available. Such systems may or may not issue a system blocking signal to apprise the originating user of the access delay.

calling signal A call control signal transmitted over a circuit to indicate that a connection is desired.

call on my dime A service feature whereby you arrange for a personal four-digit number that you give to preferred customers and business associates. Whenever or from wherever they call, the charge is billed to you. Bell Canada's Call-Me service works like this: callers dial zero and simply key in the four digit code.

call pickup A service feature of some switching systems that enables a user to dial a predetermined code and thereby answer incoming calls directed to another user in the same preselected call group. This feature is useful if you have a plain telephone set with no special buttons and want to answer calls for coworkers without having to get up from your desk and walk over to their phone to pick it up.

call processing **1.** The sequence of operations performed by a switching system from the acceptance of an incoming call through the final disposition of the call. **2.** The end-to-end sequence of operations performed by a network from the instant a call attempt is initiated until the instant the call release is completed. **3.** In data transmission, the operations required to complete all three phases of an information transfer transaction.

call progress signal A call control signal transmitted by the called data circuit-terminating equipment (DCE) to the calling data terminal equipment (DTE) to report (a) the status of a call by using a positive call progress signal or (b) the reason why a connection could not be established by using a negative call progress signal.

call progress tone An audible signal returned by a network to a call originator to indicate the status of a call. Examples of call progress tones include dial tones and busy signals.

call record Recorded data pertaining to a single call, such as date, duration of call, trunk used, extension, and other management information.

call restriction A switching system service feature that prevents selected terminals from exercising one or more service features otherwise available from the switching system.

call screening A service feature that allows you to program a list of phone numbers that you absolutely never want to talk to again. This is not very useful for most businesses; but some residential customers, home-based businesses and non-profits use this service to get rid of crank callers. Screened callers hear a polite message telling them you're not taking their calls. Period. You can also add the number that just called you. You don't have to know the number. Just press a special code.

call tracing A procedure that permits an entitled user to be informed about the routing of data for an established connection, identifying the entire route from the origin to the destination. There are two types of call tracing. Permanent call tracing permits tracing of all calls. On-demand call tracing permits tracing, upon request, of a specific call. The called party dials a designated code immediately after the call to be traced is disconnected, *i.e.,* before another call is received or placed.

call transfer A switching system service feature that allows the calling or called user to instruct the local switching equipment or switch attendant to transfer an existing call to another terminal. Call transfer may be available on a call-by-call basis or on a semipermanent basis.

call waiting A service feature that informs you when someone is trying to reach you while you are on the phone. While on the phone, a beep tone alerts you to another call. If you press hookflash quickly (or, in some parts of the country, dial a short code), you put the first call on hold and can speak to the incoming caller. To return to the previous caller, you just hookflash or press the code again. If you have caller ID service, you can link it with call waiting and see at a glance who's trying to reach you. You decide whether you want to take the call.

camp-on See **automatic callback, called-party camp-on.**

camp-on busy signal A signal that informs a busy telephone user that another call originator is waiting for a connection.

camp-on-with-recall A camp-on with the release of the call-originator terminal until the called-receiver terminal becomes free. The call originator can thus establish other calls until the recall signal is obtained, rather than simply wait until the call-receiver line is available.

carrier See **common carrier.**

carrier (signal) A wave or an unmodulated emission suitable for modulation by an information-bearing signal. The carrier is usually a sinusoidal wave or a uniform or predictable series of pulses. The original form is modified by modulating characteristics of the carrier, such as its frequency, amplitude, or phase.

carrier frequency The nominal frequency of a carrier wave. In frequency modulation, the center frequency.

carrier identification code (CIC) A four-digit number that customers use to reach a particular long-distance company other than the pre-subscribed carrier for that particular phone line. The number format, 101-XXXX, is dialed first, yielding access to the dialed carrier's network. Callers thus "dial around" the primary carrier, perhaps to get a better rate. The number format used to be 10-XXX until more CICs were necessary. See also **dial around.**

carrier sense In a local area network, an ongoing activity of a data station to detect whether another station is transmitting.

carrier sense multiple access (CSMA) A network control scheme in which a node verifies the absence of other traffic before transmitting. It is found in shared, or broadcast, networks such as Ethernet LANs, where multiple nodes connect to the network and such a scheme is needed to prevent data collisions.

carrier sense multiple access with collision avoidance (CSMA/CA) A network (Ethernet LAN) control protocol designed to allow an orderly flow of data traffic. Collisions are avoided as follows: (a) a carrier sensing scheme is used where the transmitting node listens for other traffic on the network, (b) a data station that intends to transmit sends a jam signal first, (c) after waiting a sufficient time for all stations to receive the jam signal, the data station transmits a frame, and (d) while transmitting, if the data station detects a jam signal from another station, it stops transmitting for a random time and then tries again.

carrier sense multiple access with collision detection (CSMA/CD) The most prevalent Ethernet LAN control protocol, designed to allow an orderly flow of data traffic by avoiding collisions, in which (a) a carrier sensing scheme is used where the transmitting node listens for other traffic on the network, (b) a transmitting data station that detects another signal while transmitting a frame, stops transmitting that frame, transmits a jam signal, and then waits for a random time interval before trying to send that frame again.

carrier system A multichannel telecommunications system in which a number of individual circuits (data, voice, or combination thereof) are multiplexed for transmission between nodes of a network. In carrier systems, many different forms of multiplexing may be used, such as time-division multiplexing and frequency-division multiplexing. Multiple layers of multiplexing may ultimately be performed upon a given input signal; *i.e.*, the output resulting from one stage of modulation may in turn be modulated. At a given node, specified channels, groups, supergroups, *etc.*, may be demultiplexed without affecting the others.

Category 3 The ANSI/EIA/TIA-568 designation for 100-ohm, 24 AWG (0.52 mm) unshielded twisted-pair cables and associated connecting hardware whose characteristics are specified for data transmission up to 16 Mbps.

Category 4 The ANSI/EIA/TIA-568 designation for 100-ohm, 24 AWG (0.52 mm) unshielded twisted-pair cables and associated connecting hardware whose characteristics are specified for data transmission up to 20 Mbps.

Category 5 The ANSI/EIA/TIA-568 designation for 100-ohm, 24 AWG (0.52 mm) unshielded twisted-pair cables and associated connecting hardware whose characteristics are specified for data transmission up to 100 Mbps.

CATV See **cable TV.**

CCITT Abbreviation for Consultative Committee for International Telegraph and Telephone; a predecessor of the **International Telecommunications Union, Telecommunications Services (ITU-T)**. The ITU-T is a United Nations agency based in Geneva that sets international standards and protocols that enable diverse individual national telecommunications systems to function in concert with one another, resulting in a functional worldwide network.

CDMA See **code-division multiple access.**

CDPD See **cellular digital packet data.**

cell **1.** In data communications, a cell typically refers to the 53-byte frame used in ATM networks. It is a relatively short, fixed-length data packet. **2.** In mobile telephony, a cell is the geographical area covered by one of the telephone transmitters supporting the cellular network. Cell sites are connected to cellular switching offices, which then connect to the wireline telephone network. The size of a mobile-telephony cell can vary greatly, from as little as three miles to as much as 20 miles in diameter, depending on the transmitter used and the type of terrain (city, rural, flat, mountainous) in the cell. Most cellular telephone companies have “roaming” agreements with neighboring cellular carriers, so that when one of their customer leaves the area covered by that carrier's cells, the call is relayed to a next-nearest cell belonging to an adjacent carrier.

cell relay A statistically multiplexed interface protocol for packet switched data communications that uses fixed-length packets, *i.e.,* cells, to transport data. Cell relay transmission rates usually are between 56 Kbps and 1.544 Mbps, *i.e.,* the data rate of a DO signal. Cell relay protocols (a) have neither flow control nor error correction capability, (b) are information-content independent, and (c) correspond only to layers one and two of the SO Open Systems Interconnection—Reference Model. Cell relay systems enclose variable-length user packets in fixed-length packets, *i.e.,* cells, that add addressing and verification information. Frame length is fixed in hardware, based on time delay and user packet-length considerations. One user data message may be segmented over many cells. Cell relay is an implementation of fast packet technology that is used in (a) connection-oriented broadband integrated services digital networks (B-ISDN) and (b) connectionless IEEE 802.6, switched multi-megabit data service (SMDS). Cell relay is suitable for time-sensitive traffic such as voice and video.

cell switching **1.** Refers to the method used to "switch" cellular telephone calls from one cellular switching office (also called an MTSO or Mobile Telephone Switching Office) to another. The mobile telephones used by pedestrians, cars and buses are relatively low-powered and constantly in motion. This combination means that cellular telephone companies must build enough switching offices so that as the caller and receiver move, the call can be relayed to the next-nearest switching office. At any given moment, more than one switching office might be picking up the signal from a single call. Each cellular switching office receiving signals from a call constantly measures the strength of a signal and, when it begins to fade, relays the call to the switching office that is nearer the two callers. If either of the callers moves beyond the geographic region covered by their carriers' wireless system, the call will be disconnected, or dropped. **2.** In data communications, the routing of data packets referred to as *cells*. See also **cell** and **packet switching**.

cells on wheels Mobile cellular towers that are used temporarily until a permanent tower is operational.

cellular digital packet data (CDPD) A wireless data transmission technology that sends packetized data over the analog cellular network during breaks in voice conversations or when voice channels are otherwise free. Although transmission speeds can be as high as 19.2 Kbps, in practice 9.6 Kbps is the average, dropping as low as 2 Kbps during peak usage periods. The main advantage of this technology is that it does not require a new dedicated network for transmitting data. The established cellular network is used to its fullest capacity by using idle channels. Even a break of a few seconds during voice calls is adequate to transmit packets of data. Personal digital assistants (PDAs) use this technology to send and receive data, and many cellular phones now offer this technology built in.

cellular mobile A mobile communications system that uses a combination of radio transmission and conventional telephone switching to permit telephone communication to and from mobile users within a specified area. In cellular mobile systems, large geographical areas are segmented into many smaller areas, or cells, each of which has its own radio transmitters and receivers and a single controller interconnected with the public switched telephone network. Also called *cellular radio* and *cellular telephone.*

cellular mobile telephone system (CMTS) The original analog cellular system operating in the 800 MHZ range, consisting of multiple radio transceiver sites, called cells, with overlapping coverage. The cells are connected to one another and to the wired telephone network through a mobile telephone switching office (MTSO).

cellular telephone See **cellular mobile.**

centralized attendant services (CAS) A function of a usually centrally located attendant console that permits the control of multiple switches, some of which may be geographically remote.

centralized automatic message accounting (CAMA) An automatic message accounting system that serves more than one switch from a central location.

central office A common carrier switching center in which trunks and loops are terminated and switched. This is a local phone company facility where all phone lines in the surrounding geographic area terminate. From this point the calls are *switched*, or routed, to another local phone connected to the same central office, to another central office serving a more distant area, or to the long-distance network. Also called *exchange, local central office, local exchange, local office, switching center, switching exchange, telephone exchange.*

central office code The first three digits of the local telephone number, preceded, if necessary, by the three area code digits. A central office may be assigned one or more central office codes in its geographic area. Also called *exchange.*

Centrex A central office based service provided by the local phone company, which offers PBX-like enhanced services such as conference calling, call forwarding, call hold, intercom, and caller ID. Centrex thus turns simple single-line phones into a virtual phone system without incurring the expense of buying the hardware.

channel **1.** A connection between initiating and terminating nodes of a circuit. **2.** A single path provided by a transmission medium via either (a) physical separation, such as by multipair cable or (b) electrical separation, such as by frequency- or time-division multiplexing. Multiplexing allows multiple channels to be carried over the same physical circuit.

channel bank A terminal that performs the first step of modulation by multiplexing a group of channels into a higher bandwidth analog channel or higher bit-rate digital channel and, at the receiving end, demultiplexes these aggregated signals back into individual channels. A channel bank is used to connect a group of analog devices or an analog PBX to the digital network (T1).

channel service unit (CSU) Usually the term CSU is used in combination with DSU, becoming: CSU/DSU. It stands for Channel Service Unit/Digital (or Data) Service Unit. A CSU/DSU is a device that converts frames of digital data back and forth between formats used by a LAN and a WAN. The CSU works in conjunction with the WAN circuit, both receiving and transmitting data between the subscriber and the service provider. The DSU converts certain input and output from the LAN and the time-division multiplexed frames on a T1 line and supplies timing to each user port, converting incoming data signals to the form required over a high-speed line provided by the telco. Essentially the DSU is watching for timing errors in digital data transmissions. The CSU can perform tests for the phone company, protect both telco and user equipment from lightning strikes and other electrical problems, and store usage statistics. For example, customers with T1 and fractional T1 lines will have a CSU/DSU on their own premises (CPE--customer premises equipment), and the phone company will have a CSU/DSU on its end of the network. Although a few CSUs and DSUs are manufactured and sold separately, in most instances they are combined in a single unit.

choppiness This is a less-than-optimum circumstance in which a caller's words are intermittently cut off, creating gaps in the voice transmission. This is usually the result of packet loss when transmitting voice over a packet-switched data network. Choppiness makes it difficult or impossible to have a normal conversation.

churn Industry slang for the act of subscribers changing service providers. The service provider must replace lost customers with new ones, usually from a competing service provider, who in turn must do the same. All these customers shuffling in and out amounts to churn.

circuit This term has several definitions. The most generally accepted are as follows: **1.** The complete path between two terminals over which one-way or two-way communications may be provided. **2.** An electronic path between two or more points, capable of providing a number of channels. **3.** A number of conductors connected together for the purpose of carrying an electrical current. **4.** An electronic closed-loop path among two or more points used for signal transfer.

circuit switching A method of creating a physical, dedicated connection between two or more telephones and/or data devices. POTS (regular voice) telephone service is *circuit switched.* A circuit-switched connection remains open and usable only between the parties involved, until one of the parties disconnects. Circuit-switched lines are never dedicated to particular users *when they are open.* As soon as the parties hang up, that line is available for the next calling parties to use. The other two main types of switching are *message switching* and *packet switching*. Circuit switching differs from message switching because it is interactive. Message switching involves sending a message about a call from one point on the network to another—it is sent from one place and received in another. Reciprocal action is not necessary. Packet switching is a technology initially designed for sending and receiving data only. It is like circuit switching in that no packet-switched connection is kept open on a dedicated basis. Separate users can send packet-switched data at any time. Switches in the network read the information contained in the header message of each individual packet and route that packet accordingly. The next packet, whether going to a different destination or not, is handled the same way. The packets are reassembled into the complete data transmission before delivery to the intended recipient, usually at the last switching station on the network before the recipient's data-collection device.

Citizens Band (CB) A low-power (four watts) radio spectrum divided into two frequency bands of 26.965-27.225 MHz and 462.55-469.95 MHz.

CLASS See **custom local area signaling services.**

classmark A designator used to describe the service feature privileges, restrictions, and circuit characteristics for lines or trunks that access a switch. Examples of classmarks include precedence level, conference privilege, security level, and zone restriction. Also called *class-of-service mark.*

class of office A ranking, assigned to each switching center in a communications network, determined by the center switching functions, interrelationships with other offices, and transmission requirements:

class 1 regional office (IXC)
class 2 sectional office (IXC)
class 3 primary office (IXC)
class 4 toll center (IXC point of presence)
class 5 end office (RBOC/LEC)

class of service **1.** A designation assigned to describe the service treatment and privileges given to a particular terminal. **2.** A subgrouping of telephone users for the purpose of rate distinction. Examples of class of service subgrouping include distinguishing between (a) individual and party lines, (b) Government and non-Government lines, (c) those permitted to make unrestricted international dialed calls and those not so permitted, (d) business, residence, and coin-operated, (e) flat rate and message rate, and (f) restricted and extended area service. **3.** A category of data transmission provided by a public data network in which the data signaling rate, the terminal operating mode, and the code structure are standardized. Class of service is defined in CCITT Recommendation X.1.

Also called *user service class.*

class-of-service mark See **classmark**.

clear channel **1.** In radio broadcasting, a frequency assigned for the exclusive use of one entity. **2.** In networking, a signal path that provides its full bandwidth for a user's service. No control or signaling is performed (or needed) on this path.

CLEC See **competitive local exchange carrier**.

client-server architecture Any network-based software system that uses client software to request a specific service, and corresponding server software to provide the service from another computer on the network. The client is usually a PC on a LAN, and the server is often another PC on the LAN, perhaps more powerful or with a bigger hard drive. The server acts as a central storage facility for the organization's files, which need to be accessed, worked upon, and updated by multiple clients on the network. Everyone has access to the same up-to-date data. The server can also hold software application programs, such as word processing programs, making updates to the software much easier—only one computer needs updating instead of all the client computers. This arrangement also makes back-ups and security easier because only one central computer is involved.

clipping In telephony, the loss of the initial or final parts of a word, words, or syllable, usually caused by the faulty operation of voice-actuated devices.

clock rate The rate at which a clock issues timing pulses. Clock rates are usually expressed in pulses per second, such as 4.96 Mpps (megapulses per second).

cloning Whenever a cellular phone makes an outgoing call, it transmits both its phone number and the electronic serial number (ESN) unique to that phone. People with scanners can intercept this information and program it into another cellular phone, thus illegally cloning the original phone. The legal owner of the original phone will eventually receive a huge bill for the calls from the cloned phone.

closed captioning In broadcast and cable television, the insertion of information that may be decoded and displayed on the screen as written words corresponding to those being spoken and transmitted via the conventional audio subcarrier. Closed captioning, developed for the hearing impaired, requires a special decoder, which may be external to, or built into, the television receiver. Closed captioning is mandated by the Americans with Disabilities Act of 1990.

C.O. See **central office**.

coaxial cable (coax) A cable consisting of a center conductor surrounded by an insulating material and a concentric outer conductor. Coaxial cable is used primarily for wideband, video, CATV, or RF applications because it is capable of carrying large amounts of data.

codec Acronym for **coder-decoder**. A conversion device that translates analog electrical signals into bipolar digital signals suitable for transmission over telephone networks.

code division multiple access (CDMA) The wireless switching technology that PCS (Personal Communications Service) providers are expected to use in their communications networks. CDMA can also be referred to as a "spread spectrum" technology used in digital cellular telephones. The term "spread spectrum" refers to the fact that transmissions, which are first transformed into data packets that contain identification information, are literally "spread" over a wide range of frequencies. Spreading transmissions in this way allows a specified range of frequencies to carry up to ten distinct transmissions simultaneously, whereas a comparable transmission channel using analog technology can carry only one transmission.

There is debate as to whether this technology is an improvement over traditional cellular telephony. The appeal of CDMA spread spectrum technology is largely twofold: first, it carries about ten times more traffic than a comparable analog line and thus uses allocated spectrum more efficiently; second, CDMA is one of three major wireless transmission technologies. The others are GSM (Global Standard for Mobile communications) and TDMA (time division multiple access). In CDMA, the data being relayed is transformed to a digital format and spreads out to occupy the entire available bandwidth. When more than one call needs to use the same route, each is assigned a unique sequence code to the bits of speech, essentially putting them in digital code. This unique digital code can only be received by the specific wireless phone it is intended for, thus making CDMA transmissions more secure than regular cellular transmissions, which can be overheard by others. Once this digital code is received by the intended party, it is translated back into speech. A company called Qualcomm Inc., San Diego, developed CDMA technology.

collision In a data transmission system, the situation that occurs when two or more demands are made simultaneously on equipment that can handle only one at any given instant. For example, a collision will occur when two computers on a LAN try to send data to each other at the same instant.

co-location (colocation) The placement of network equipment of one company in the offices or buildings of the local phone company for the purpose of connecting to the local network and facilities. The co-locating company can be a competing local telephone company, a long-distance carrier, an Internet service provider (ISP) or an end-user company.

combined distribution frame (CDF) A distribution frame that combines the functions of main and intermediate distribution frames and contains both vertical and horizontal terminating blocks. The vertical blocks are used to terminate the permanent outside lines entering the station (central office). Horizontal blocks are used to terminate inside plant equipment. This arrangement permits the connection of any outside line with any desired terminal equipment. These connections are made either with twisted pair wire, normally referred to as jumper wire, or with optical fiber cables, normally referred to as jumper cables. In technical control facilities, the vertical side may be used to terminate equipment as well as outside lines. The horizontal side is then used for jackfields and battery terminations.

command frame In data transmission, a frame, containing a command, transmitted by a primary station.

common carrier A telecommunications company that holds itself out to the public for hire to provide communications transmission services. In the United States, such companies are usually subject to regulation by federal and state regulatory commissions and must file tariffs that outline their service offerings and fee structures. Also called *carrier, commercial carrier, communications common carrier.*

common-channel interoffice signaling (CCIS) In multichannel switched networks, a method of transmitting all signaling information for a group of trunks by encoding it and transmitting it over a separate channel using time-division digital techniques.

common-channel signaling In a multichannel communications system, signaling in which one channel in each link is used for signaling to control, account for, and manage traffic on all channels of the link. The channel used for common-channel signaling does not carry user information.

common control switching arrangement (CCSA) An arrangement in which switching for a private network is provided by one or more common control switching systems. The switching systems may be shared by several private networks and also may be shared with the public telephone networks.

communications **1.** Information transfer, among users or processes, according to agreed conventions (telephone, Morse code, smoke signals, signal flags, hand signals, English language, etc.). **2.** The branch of technology concerned with the representation, transfer, interpretation, and processing of data among persons, places, and machines. The meaning assigned to the data must be preserved during these operations.

competitive access provider (CAP) A company that provides exchange access services in competition with an established U.S. telephone local exchange carrier. A CAP usually provides a direct, discount link between major telephone customers and their long-distance carrier, bypassing the local carrier.

competitive local exchange carrier (CLEC) A carrier that provides full local service in competition with a dominant local phone company, called the *incumbent local exchange carrier (ILEC).*

complete the call A service feature that works in conjunction with the phone company's directory assistance service. Callers to directory assistance hear a message offering to dial the number for them, free of charge as a courtesy, or for a stated fee.

computer network **1.** A network of data processing nodes that are interconnected for the purpose of data communication. **2.** A communications network in which the end instruments are computers.

computer network operating system (NOS) A specialized operating system designed for computer networking on minicomputers and microcomputers in a local networking area / campus area network. An NOS is usually designed to run on existing software designed for that computer and may require interface hardware for the workstation and server.

computer telephony The convergence of computer and telecommunications technologies. Microchips and computers allow for all kinds of sophisticated automated capabilities to be added to the basic telephone. Fax-on-demand, interactive voice response, and videoconferencing are all the results of computer telephony. Also known as *computer-telephone integration (CTI)*.

concentrator **1.** In data transmission, a functional unit that permits a common path to handle more data sources than there are channels currently available within the path. A concentrator usually provides communication capability between many low-speed channels (usually asynchronous) and one or more high-speed channels (usually synchronous). Different speeds, codes, and protocols can be accommodated on the low-speed side. The low-speed channels usually operate in contention and require buffering. **2.** A device that connects a number of circuits, which are not all used at once, to a smaller group of circuits for economy.

conditioned circuit A communications circuit optimized to obtain desired characteristics for voice or data transmission.

conditioned loop A loop that uses conditioning equipment to obtain the desired line characteristics for voice or data transmission. The conditioning equipment is used to improve the amplitude-vs.-frequency characteristics of the circuit and to match impedance.

conference bridge See **conference call.**

conference call A feature that allows a call to be established among three or more stations in such a manner that each of the stations is able to communicate with all the other stations. It can be a service provided by a telephone company, with all parties dialing into one number and then being connected via a conference bridge, or this capability can be built into the end user's telephone system.

contention **1.** A condition that arises when two or more data stations attempt to transmit at the same time over a shared channel, or when two data stations attempt to transmit at the same time in two-way alternate communication. A contention can occur in data communications when no station is designated a master station. In contention, each station must monitor the signals and wait for a quiescent condition before initiating a bid for master status. **2.** Competition by users of a system for use of the same facility at the same time.

cordless telephone A combination of wired and wireless technology that comes in two parts: The base station that connects to the phone jack and the wired telephone network, and the portable handset with a built-in radio transceiver that communicates with the base station. Two frequency options are currently available: 46/49 MHz with optimum ranges from 300 to 1,000 feet, and 900 MHz with ranges up to a half mile. Actual ranges are shorter due to interference and obstructions.

country code In international direct telephone dialing, a code that consists of 1-, 2-, or 3-digit numbers in which the first digit designates the region and succeeding digits, if any, designate the country. A list of these codes, plus codes for the major cities located within these countries, can usually be found in the front of the phone directory. When dialing from the United States, the digits "011" must be dialed first for all international direct-dialed calls.

COW See **cells on wheels.**

CPE See **customer premises equipment.**

CPNI See **customer proprietary network information.**

cramming The practice of including unauthorized, deceptive, or misleading charges on the telephone bill for products or services that were not ordered or approved by the subscriber. These charges can originate from the local exchange carrier (LEC) itself or from other entities for which the LEC performs billing and collection services. The LEC often serves as the collection agent for other entities such as long-distance carriers or third-party billers. This arrangement makes it possible for unscrupulous operators to take advantage of the LEC's billing services and "cram" the bills with bogus charges. With so many new telephone companies, particularly long-distance resellers, it is difficult for LECs to ascertain the legitimacy of every company or its charges.

crosstalk **1.** Undesired capacitive, inductive, or conductive coupling from one circuit, part of a circuit, or channel, to another. **2.** Any phenomenon by which a signal transmitted on one circuit or channel of a transmission system creates an undesired effect in another circuit or channel. In telephony, crosstalk is usually distinguishable as speech or signaling tones.

CSMA See **carrier sense multiple access.**

CSMA/CA See **carrier sense multiple access with collision avoidance.**

CSMA/CD See **carrier sense multiple access with collision detection.**

CSU See **channel service unit.**

customer premises equipment (CPE) Communications equipment and inside wiring located at a subscriber's premises and connected with a carrier's communication channel(s) at the demarcation point (demarc). The demarc is a point established in a building or complex to separate customer equipment from telephone company equipment. Also called *customer provided equipment.*

customer proprietary network information (CPNI) This is information that a carrier collects about its customers by virtue of its relationship as a service provider, relating to the quantity, technical configuration, type, destination and amount of use of a telecommunications service subscribed to by the customer. To safeguard the consumers' privacy, Section 222 of the Telecom Act of 1996 and subsequent FCC rulings limit the amount of this information that can be used to market other services to the consumer.

custom local area signaling service (CLASS) One of an identified group of network-provided enhanced services. A CLASS group for a given network usually includes several enhanced service offerings, such as incoming-call identification, call trace, call blocking, automatic return of the most recent incoming call, call redial, and selective forwarding and programming to permit distinctive ringing for incoming calls.

cutover The physical changing of circuits or lines from one configuration to another.

D

data Representation of facts, concepts, or instructions in a formalized manner suitable for communication, interpretation, or processing by humans or by automatic means. Any representations such as characters or analog quantities to which meaning is or might be assigned. In data communications, most data is transmitted in a digital format.

data bank **1.** A set of data related to a given subject and organized in such a way that it can be consulted by users. **2.** A data repository accessible by local and remote users. A data bank may contain information on single or multiple subjects, may be organized in any rational manner, may contain more than one database, and may be geographically distributed. More than one data bank may be required to build a comprehensive database.

database **1.** A set of data that is required for a specific purpose or is fundamental to a system, project, enterprise, or business. A database may consist of one or more data banks and be geographically distributed among several repositories. **2.** A formally structured collection of data. In automated information systems, the database is manipulated using a database management system.

data circuit-terminating equipment (DCE) The equipment located at the network end of the line that performs signal conversion, coding, and other functions between the data terminal equipment (DTE) and the phone line. DCE may be a separate device or an integral part of the DTE. Network interface cards (NICs), DSUs, CSUs, modems and routers are all DCE. A computer is DTE. Also called *data communications equipment.*

data communication The transfer of information between functional units by means of data transmission according to a protocol, such as Ethernet or frame relay. Data are transferred from one or more sources to one or more sinks (destinations) over one or more data links.

data compaction The reduction of the number of data elements, bandwidth, cost, and time for the generation, transmission, and storage of data without loss of information. It is accomplished by eliminating unnecessary redundancy, removing irrelevancy, or using special coding. Examples of data compaction methods are the use of fixed-tolerance bands, variable-tolerance bands, slope-keypoints, sample changes, curve patterns, curve fitting, variable-precision coding, frequency analysis, and probability analysis. Simply squeezing noncompacted data into a smaller space--such as by increasing packing density or transferring data on punched cards onto magnetic tape--is not data compaction. Whereas data compaction reduces the amount of data used to represent a given amount of information, data compression does not.

data compression **1.** Increasing the amount of data that can be stored in a given domain, such as space, time, or frequency, or contained in a given message length. **2.** Reducing the amount of storage space required to store a given amount of data, or reducing the length of message required to transfer a given amount of information. Data compression may be accomplished by simply squeezing a given amount of data into a smaller space, for example, by increasing packing density or by transferring data on punched cards onto magnetic tape. Data compression does not reduce the amount of data used to represent a given amount of information, whereas data compaction does. Both data compression and data compaction result in the use of fewer data elements for a given amount of information.

data corruption The violation of data integrity. Also called *data contamination.*

data country code A 3-digit numerical country identifier that is part of the 14-digit network terminal numbering plan. The data country code prescribed numerical designation further constitutes a segment of the overall 14-digit X.121 numbering plan for a CCITT X.25 network.

datagram In packet switching, a self-contained, independent packet that contains information sufficient for routing from the originating data terminal equipment (DTE) to the destination DTE without relying on prior exchanges between the equipment and the network. Unlike virtual call service, when datagrams are sent, there are no call establishment or clearing procedures. Thus, the network may not be able to provide protection against loss, duplication, or misdelivery. The Internet uses datagrams. See also **Internet protocol (IP)**.

data link **1.** The means of connecting one location to another for the purpose of transmitting and receiving data. **2.** An assembly, consisting of parts of two data terminal equipments (DTEs) and the interconnecting data circuit, that is controlled by a link protocol that enables data to be transferred from a data source to a data sink.

data network identification code (DNIC) In the CCITT International X.121 format, the first four digits of the international data number, the three digits that may represent the data country code, and the 1-digit network code, *i.e.*, the network digit.

data numbering plan area (DNPA) In the U.S. implementation of a CCITT

X.25 network, the first three digits of a network terminal number (NTN). The 10-digit NTN is the specific addressing information for an end-point terminal in an X.25 network.

data rate The aggregate rate at which data pass a point in the transmission path of a data transmission system, usually expressed in bits per second. Also called the *data throughput, data transfer rate, transmission speed* and *data signaling rate*. The transmission rates of various channels/media are compared in figure D-1.

data service unit (DSU) A conversion device that translates the digital output from a computer (square waves) into bipolar digital pulses that are suitable for transmission over a digital communications network. It is usually combined with a **channel service unit (CSU)**.

data signaling rate (DSR) See **data rate**.

data stream A sequence of digitally encoded signals used to represent information in transmission.

data switching exchange (DSE) The equipment installed at a single location to perform switching functions such as circuit switching, message switching, and packet switching.

data terminal equipment (DTE) An end instrument that converts user information into signals for transmission or reconverts the received signals into user information. Computers are DTEs. The DTE connects to the network via data circuit-terminating equipment (DCE), such as modems or DSU/CSUs.

data transfer rate The average number of bits, characters, or blocks per unit time passing between corresponding equipment in a data transmission system. See **data rate**.

data transfer request signal A call control signal transmitted by the data circuit-terminating equipment (DCE) to the data terminal equipment (DTE) indicating that a request signal, originated by a distant DTE, has been received from a distant DCE to exchange data with the station.

data transmission The sending of data from one place to another by means of signals over a channel.

data transmission circuit The transmission media and the intervening equipment used for the transfer of data between data terminal equipments (DTEs). A data transmission circuit includes any required signal conversion equipment such as modems or multiplexers. A data transmission circuit may transfer information in (a) one direction only, (b) either direction but one way at a time, or (c) both directions simultaneously.

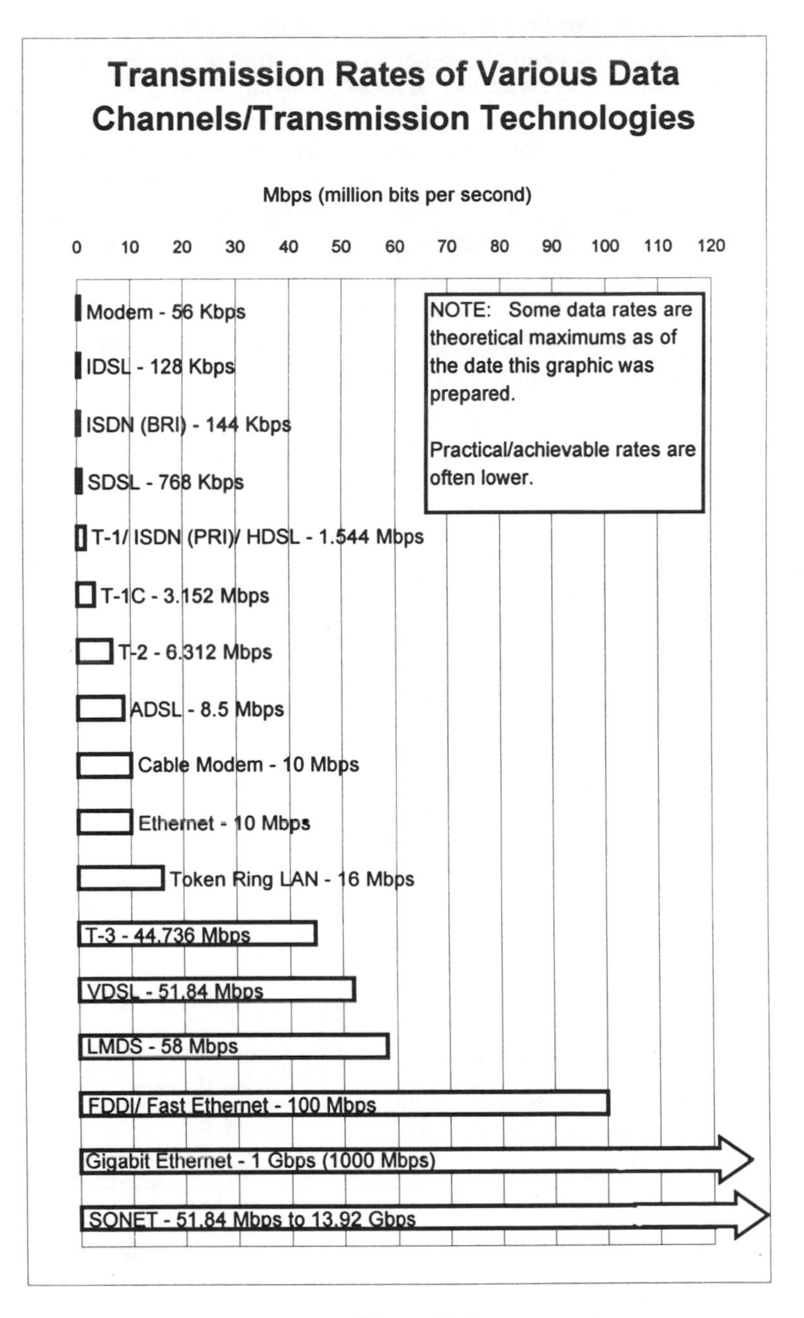
Transmission Rates of Various Data Channels/Transmission Technologies
Mbps (million bits per second)
0 10 20 30 40 50 60 70 80 90 100 110 120
Modem - 56 Kbps
IDSL - 128 Kbps
ISDN (BRI) - 144 Kbps
SDSL - 768 Kbps
T-1/ ISDN (PRI)/ HDSL - 1.544 Mbps
T-1C - 3.152 Mbps
T-2 - 6.312 Mbps
ADSL - 8.5 Mbps
Cable Modem - 10 Mbps
Ethernet - 10 Mbps
Token Ring LAN - 16 Mbps
T-3 - 44.736 Mbps
VDSL - 51.84 Mbps
LMDS - 58 Mbps
FDDI/ Fast Ethernet - 100 Mbps
Gigabit Ethernet - 1 Gbps (1000 Mbps)
SONET - 51.84 Mbps to 13.92 Gbps
NOTE: Some data rates are theoretical maximums as of the date this graphic was prepared.
Practical/achievable rates are often lower.

Figure D-1

dB Abbreviation for **decibel(s).** A measure of intensity, such as the loudness of sound, expressed as one tenth of the common logarithm of the ratio of relative powers, equal to 0.1 B (bel). The decibel is the conventional relative power ratio, rather than the bel, for expressing relative powers because the decibel is smaller and therefore more convenient than the bel. The dB is used rather than arithmetic ratios or percentages because when circuits are connected in tandem, expressions of power level, in dB, may be arithmetically added and subtracted. For example, in an optical link, if a known amount of optical power, in dBm, is launched into a fiber, and the losses, in dB, of each component (*e.g.*, connectors, splices, and lengths of fiber) are known, the overall link loss may be quickly calculated with simple addition and subtraction.

DCE See **data circuit-terminating equipment.**

DCE clear signal A call control signal transmitted by data circuit-terminating equipment (DCE) to indicate that it is clearing the associated circuit after a call is finished.

DCE waiting signal A call control signal at the interface of the data-circuit terminating-equipment/data-terminal-equipment (DCE/DTE) that indicates the DCE is ready for another event in the call establishment procedure.

D channel In ISDN, the 16-Kbps segment of a 144-Kbps, full-duplex subscriber service channel that is subdivided into 2B+D channels, *i.e.*, into two 64-Kbps clear channels and one 16-Kbps channel for the ISDN basic rate. The D channel is usually used for out-of-band signaling. The two 64-Kbps clear channels are used for subscriber voice and data services. The D-channel specifications are addressed in the CCITT Recommendation for the Integrated Services Digital Network (ISDN). The D-channel may be 64 Kbps for the primary rate ISDN service.

dedicated circuit A circuit designated for exclusive use by specified users.

demarcation point (demarc) That point at which operational control or ownership of communications facilities changes from the customer to the network service provider. The demarcation point is usually the interface point between customer-premises equipment (CPE) and the telephone company's equipment. Also called *network terminating interface.*

demultiplexing The separation of two or more channels previously multiplexed; *i.e.*, the reverse of multiplexing.

dense wave division multiplexing (DWDM) See **wavelength division multiplexing.**

dial around A method that allows callers to dial a code and use any long-distance carrier for a single call. Every long-distance carrier has a four-digit access code, called the **carrier identification code (CIC)**, in the format 101-XXXX. Dialing this number (before dialing the other party's number) allows callers to use a specific carrier from any location, dialing around the primary carrier preselected for the particular phone line currently in use.

dial-a-porn Refers to adult services offered over pay-per-call (900) lines, such as live one-on-one "romance" or "fantasy" lines. Services targeted to adults that do not use indecent language are legal. Indecent language is defined by the FCC as "the description or depiction of sexual or excretory activities or organs in a patently offensive manner as measured by contemporary standards for the telephone medium."

dialing In a communications system, using a device that generates signals for selecting and establishing connections. For standard telephones, using the rotary dial or the keypad to input the desired telephone number.

dialing parity A term created by the Telecommunications Act of 1996, meaning that customers of new competitive telephone companies are not required to dial special access codes in addition to the regular phone number. All dialing patterns are the same, regardless of the carrier used.

dial pulsing Pulsing in which a direct-current pulse train is produced by interrupting a steady signal according to a fixed or formatted code for each digit and at a standard pulse repetition rate. Dial pulsing originated with rotary mechanical devices integrated into telephone instruments, for the purpose of signaling. Subsequent applications use electronic circuits to generate dial pulses.

dial service assistance (DSA) A network-provided service feature, associated with the switching center equipment, in which services, such as directory assistance, call interception, random conferencing, and precedence calling assistance, are rendered by an attendant.

dial signaling Signaling in which dual tone multifrequency (DTMF) signals or pulse trains are transmitted to a switching center. Rotary dials produce pulse trains. Keypads may produce either DTMF signals or pulse trains. Dial signaling traditionally refers to pulse trains only.

dial through A technique, applicable to access circuits, that permits an outgoing routine call to be dialed by the PBX user after the PBX attendant has established the initial connection.

dial tone A tone employed in a dial telephone system to indicate to the calling party that the equipment is ready to receive dial or tone pulses.

dial-up **1.** A service feature in which a user initiates service on a previously arranged trunk or transfers, without human intervention, from an active trunk to a standby trunk. **2.** A service feature that allows a computer terminal to use telephone systems to initiate and effect communications with other computers.

DID See **direct inward dialing.**

differentiated services (DiffServ) A networking protocol that can discriminate among different quality of service (QoS) profiles and route data traffic according to its assigned priority. This protocol enables routers throughout the network to recognize different packets as carrying voice, file transfer, or Web data packets, prioritizing them as a "platinum," "gold," "silver," or "bronze" class of service (QoS), and routing them accordingly.

digital Characterized by discrete states, *i.e.*, the presence or absence of a signal. Digital data is represented by discrete values or conditions, as opposed to analog data, which is characterized by continuously changing conditions (sine wave). Also the representation of numbers by digits, perhaps with special characters and the "space" character, such as the "1s" and "0s" of the binary digital system used by computers and data communications. In data transmissions over copper wire, for example, the discrete states are the presence or absence of an electrical voltage (3 volts) at set timing intervals. The presence of voltage represents a "1," and no voltage represents a "0."

digital cellular A superior transmission technology to analog cellular, capable of carrying voice, data and video. Allows for clearer voice quality than analog cellular, and enables enhanced services such as caller ID, voice mail and messaging. See also **Personal Communications Services (PCS), time division multiple access (TDMA), code division multiple access (CDMA)** and **Global System for Mobile communications (GSM).**

digital loop carrier (DLC) The equipment, including lines, that is used for digital multiplexing of telephone circuits, and that is provided by the network as part of the subscriber access.

digital multiplexer A device for combining several digital signals into an aggregate bit stream. Digital multiplexing may be implemented by interleaving bits, in rotation, from several digital bit streams either with or without the addition of extra framing, control, or error detection bits.

digital signal 0-4 (DS 0-4) All are speed classifications of digital signal (or data service), levels 0 through 4. These terms were originated by AT&T to describe the line speeds or digital data carrying capacity of various lines and trunks available for lease in its network. Since then, various levels of DS service have been standardized. Using various DS level services, bits of data on a communication line are surrounded by "protocol bits" that direct the transmission. DS0 represents a single voice conversation transmitted as a single digital data stream. It operates at 64 Kbps. DO represents 24 conversations multiplexed together, operating at a transmission rate of 1.544 Mbps over a T1 line. DS1C operates at 3.152 Kbps, also on a T1 line; DS2 is carried on a T2 line and operates at 6.312 Mbps. DS3 is the equivalent of 28 T1 channels, using T3 lines, operating at 44.736 Mbps. DS3 can be used on T3 ISDN lines. DS4 is the equivalent of 168 T1 channels, using T4 lines, operating at 274.176 Mbps.

digital subscriber line (DSL) This developing technology uses ordinary copper telephone lines (which comprise most of the local telephone companies' networks) to deliver high-speed information. DSL appeals to telephone companies because, in many cases, it is faster and less expensive than replacing their existing copper-wire network with higher-speed fiber-optic lines. (Still, many phone companies, particularly long-distance carriers, do have largely fiber networks.) Theoretically, DSL can deliver information, including audio, video, and text, at speeds up to nearly 8.5 megabits per second downstream to the consumer. Telephone companies and equipment vendors are still experimenting with a variety of speed combinations to find the most effective and economical combination. DSL proponents believe that this technology will, at some point, largely replace Integrated Services Digital Network (ISDN).

The primary appeal of xDSL (the *x* denotes several possible variations) technology, designed to deliver high-bandwidth information to homes and small businesses, is that it operates over the existing copper-wire network (ordinary telephone lines) and does not require service providers to rebuild their networks using fiber-optic technology. Of nearly equal appeal is the fact that DSL technologies enable the telephone company to "split" the voice and data signal coming in to the customer's premises, giving them in effect two separate lines: one to carry voice signals to the telephone, another to carry the data signals to a computer or other digital device. With this capability, customers could leave their computer connected to the Internet constantly (a state called *always on* in the industry) while still having the option of making and receiving voice telephone calls. At this time, most of us use a computer modem to receive digital data from the Internet or other sources. With current modem technology, data begins its journey in digital format, is transformed by a modem in the telephone network to an analog signal, and is switched back to digital by the modem attached to one's own computer. With xDSL, the digital information is never transformed into analog format and therefore uses a much smaller portion of the bandwidth in the telephone network. Copper wires can carry a much greater amount of information in a digital format. Since DSL allows the signal to begin, be transmitted, and received all in digital format, each pair of copper wires can carry a much greater amount of data.

Most DSL technologies require the telephone company to install something called a "splitter" at the customer's premises. This could be a costly and time consuming procedure if demand for DSL turns out to be high and the telephone company has to send a service truck to each customer to perform the installation. Many companies are working on a form of DSL referred to as *splitterless DSL* or *Universal DSL* or *DSL Lite*. Using this iteration of DSL, the telephone company itself splits the signal remotely from its central office, thus eliminating the need to send a truck to the customer's home. In this case, however, many customers would have to install a filter on their own telephone to eliminate interference between the signals from the voice and digital portions of the line. There are questions among some experts as to how effective splitterless DSL will be. See also **ADSL**.

digital subscriber line access multiplexer (DSLAM) A device, usually located at the central office, where individual DSL circuits (local loops) are aggregated through time division multiplexing (TDM) and then connected via a high-speed digital link to a packet-switched network such as the Internet.

digital switching Switching in which digitized signals are switched without converting them to or from analog signals.

digital transmission The binary transmission system, so called because it makes use of two digits, "0" and "1," to encode information. The spoken voice, originally an analog signal, can be changed into digital signals by assigning a "0" or "1" to minute portions of each sound. A digital signal is much more accurate and easier to transmit and reproduce than is an analog signal. For example, when an analog signal moves through a network as a wave that varies in frequency and amplitude, noise in the line attaches itself to the electrical signal that is emulating a person's voice. As these electrical signals are conducted down the line toward the receiver they need to be amplified periodically, and ALL of the signals, both the originals and any accumulated but unrelated electrical sounds, are magnified and passed on to the receiver. The end result can be a lot of hissing or other background noise in a line. Digital transmissions, on the other hand, pass through repeaters in order to maintain signal strength, where the characteristics of the signal—the presence or absence of a voltage representing a "1" or a "0"—are never varied, lost, or distorted. The transmission reaches its destination in exactly the same condition as it was sent, hence sound quality is much better.

direct inward dialing (DID) A service feature that allows inward-directed calls to a PBX to reach a specific PBX extension without human intervention, bypassing the attendant. For example, the main company number might be 555-1000, and its 100 extensions numbered 1001 through 1100. The person at each of these extensions may be dialed directly (i.e., 555-1025).

direct inward system access (DISA) A telephone system feature that allows a remote caller to access designated lines and features of the phone system. For example, an employee could call into the system from his home phone in order to place a long-distance call, and by entering the appropriate codes, the call is placed using the company's carrier and is charged to the company's telephone bill.

direct outward dialing (DOD) An automated PBX service feature that provides for outgoing calls to be dialed directly from the user terminal, usually by dialing a "9" to get an outside dial tone. Virtually all new PBXs have this feature.

distinctive ring A service feature that assigns multiple phone numbers to the same line or circuit. Each phone number rings with a different cadence so that the person receiving the call can tell which number was dialed. It can be used for teenagers in a household or to differentiate fax and voice calls. Switches can be purchased that recognize the different ring patterns and route the call to the appropriate device. This service goes by many names throughout the country including *RingMate, Identa Ring, Ring Master, Personalized Ring, Custom Ringing, Route-a-Call, Teen Service, Multi-Line* and *Smart Ring*. You pay a monthly fee per number, usually from $4 to $7, a lot less than a separate line.

distribution frame A piece of equipment that interconnects wires between two sources, located in central offices, PBXs, or other switching facilities. In a central office, for example, all the subscriber's wires come into the building and first connect to one side of the distribution frame. The internal C.O. wiring connects to the other side of the distribution frame. Jumper wires running in between connect the two sides, allowing for changes to be made when telephone numbers are changed, added, or dropped.

divestiture The court-ordered separation of the Bell Operating Telephone Companies from AT&T in 1983, when AT&T ceased to be the monopoly telephone company in the United States. The seven Regional Bell Operating Companies (RBOCs) were given control of local phone services while AT&T kept the long-distance market.

domain name server A server that retains the addresses and routing information for TCP/IP LAN users.

Domain Name System (DNS) The online distributed database system that (a) is used to map human-readable addresses into Internet Protocol (IP) addresses, (b) has servers throughout the Internet to implement hierarchical addressing that allows a site administrator to assign machine names and addresses, (c) supports separate mappings between mail destinations and IP addresses, and (d) uses domain names that (i) consist of a sequence of names, *i.e.*, labels, separated by periods, *i.e.*, dots, (ii) usually are used to name Internet host computers uniquely, (iii) are hierarchical, and (iv) are processed from right to left, such as the host nic.ddn.mil has a name (nic — the Network Information Center), a subdomain (ddn — the Defense Data Network), and a primary domain (mil — the MILNET).

do not disturb A service feature that lets you get work done without being constantly interrupted by a ringing phone. You can either turn this on and off at will (by pressing *78) or set up a regular schedule in advance. You can give family members or preferred clients a privileged "caller code" that lets their calls ring through anyway.

downlink A data link from a satellite or other spacecraft to a terrestrial terminal.

downstream In communications, the direction of transmission flow from the source toward the sink (destination/user).

DSL See **digital subscriber line**.

DSLAM See **digital subscriber line access multiplexer**.

DTE See **data terminal equipment.**

DTMF See **dual tone multifrequency.**

dual tone multifrequency (DTMF) The technical term describing push button or touch-tone dialing. When you touch a button on a telephone keypad, it makes a tone, which is actually a combination of two tones, one high frequency and one low frequency. Hence the name dual tone multifrequency. DTMF signals, unlike dial pulses, can pass through the entire connection to the destination user, and therefore lend themselves to various schemes for remote control after access, *i.e.*, after the connection is established, such as interactive voice response. Telephones using DTMF usually have 12 keys. Each key corresponds to a different pair of frequencies. Each pair of frequencies corresponds to one of the ten decimal digits, or to the symbols "#" or "*", the "*" being reserved for special purposes.

DWDM See **dense wave division multiplexing**.

E

E-commerce Electronic commerce. This is not a precisely defined term, but generally refers to the conduct of commerce by transmitting data over telecommunications networks, both wireline and wireless. The use of the Internet and the World Wide Web to conduct commerce is the most recognizable example. Other examples include the use of networks (LANs, MANs, WANs) to transact business with suppliers and customers, transfer funds, authorize payments or otherwise transmit data that facilitates commercial transactions.

EDI See **electronic document interchange.**

EHF See **extremely high frequency**.

eight hundred (800) service A service that allows call originators to place toll telephone calls to 800 service subscribers from specified rate areas within the North American Numbering Plan (NANP), without a charge to the call originator. The numbers 888 and 877 have been added to this service to expand available numbers. Also called *toll-free* service.

electromagnetic radiation (EMR) Radiation made up of oscillating electric and magnetic fields and propagated with the speed of light. Includes gamma radiation, X-rays, ultraviolet, visible, and infrared radiation, and radar and radio waves.

electromagnetic spectrum The range of frequencies of electromagnetic radiation from zero to infinity. The electromagnetic spectrum was, by custom and practice, formerly divided into 26 alphabetically designated bands. This usage still prevails to some degree. However, the ITU formally recognizes 12 bands, from 30 Hz to 3000 GHz. New bands, from 3 THz to 3000 THz, are under active consideration for recognition.

electronic document interchange (EDI) Standards and protocols that allow computers at two or more related organizations to exchange common documents such as invoices, purchase orders, and bills of lading. This allows affiliated enterprises, such as a distributor and its retailer customers, to place orders, remit payments, and track all transactions between companies.

electronic mail (e-mail) An electronic means for communicating primarily text, although files containing graphics, audio or video can be attached to the e-mail message (the recipient will need the appropriate software to "read" such attachments). Operations include sending, storing, processing, and receiving information transmitted over data networks such as LANs or the Internet. Messages are held in storage, usually on the service provider's computer hard drive, until called for (downloaded) by the addressee. Many feel that e-mail is the most useful feature of the Internet.

electronic switching system (ESS) **1.** A telephone switching system based on the principles of time-division multiplexing of digitized analog signals. An electronic switching system digitizes analog signals from subscribers' loops, and interconnects them by assigning the digitized signals to the appropriate time slots. It may also interconnect digital data or voice circuits. **2.** A switching system with major devices constructed of semiconductor components. A semi-electronic switching system that has reed relays or crossbar matrices, as well as semiconductor components, is also considered to be an ESS.

ELF See **extremely low frequency**.

e-mail See **electronic mail.**

emoticon Contraction of emotional icon. The use of standard punctuation symbols usually to create a face that conveys an emotion by the typist. Some of the more common emoticons:

:-) smile	:-(sad
:-] smirk	;-) wink
:-D laugh	:-O mouth agape

end user The ultimate user of a telecommunications service.

enhanced services Services provided by the telephone company over its network facilities that may be provided without filing a tariff, usually involving some computer-related feature such as formatting data or restructuring the information. Examples are optional, non-tariffed services such as caller ID, call forwarding, call waiting, and voice mail.

erlang A dimensionless unit of the average traffic intensity (occupancy) of a facility during a period of time, usually a busy hour. Erlangs, a number between 0 and 1, inclusive, is expressed as the ratio of (a) the time during which a facility is continuously or cumulatively occupied to (b) the time that the facility is available for occupancy. A unit of measurement of telephone traffic.

error checking/detection Data packets sent over a network must be checked for errors during transmission and when they are received. In public frame relay networks, for example, the network itself checks for errors, but correction of errors is handled by CPE (customer premises equipment) at the users' site. Some common examples of protocols for error checking include:

parity check: This is the simplest form of detecting errors in an asynchronous data transmission. It involves adding a single bit of data to a an ASCII-formatted series of zeroes and ones that represent a single data character. By checking the data packet and counting the number of zeroes and ones, the parity checking system determines whether it is an "odd" or "even" parity check, which simply means the number of ones in the character is a mathematically odd or even number. If the predetermined protocol is for an "even" parity check, all data characters with an odd number of ones in the packet will have a single bit representing a one added to the stream, making an even number of ones. The receiving system checks to see if the data packet contains either the agreed-upon even or odd number of ones in the bit stream. If it does not, it can request that information be resent.

vertical redundancy check (VRC): Similar to parity check; also used for asynchronous transmissions. The packets of bits representing characters of data are arranged vertically. Once again, the transmitting computer checks each data packet, counting the number of zeroes and ones, and appends the parity bit to alert the receiving computer what to expect. It is not a particularly efficient method of checking for errors—they are easily missed. Furthermore, if errors are extreme enough to make a transmission illegible, a person has to resolve the situation. A VRC will not automatically trigger a redial and resend. When longitudinal redundancy checking (LRC) is used in conjunction with VRC, it improves the likelihood of catching errors.

longitudinal redundancy checking (LRC): A more complex method of error checking which also involves counting the numbers of zeroes and ones in each character but where each block of data is appended with an LRC character. Called the block check character (BCC), this is the last character sent in a block of data. It is set at the sending end of the transmission; on the receiving end the BCC is checked and the block of data is disassembled, a BCC is set, and is compared with the BCC attached at the sending end. If both coincide, the transmission is error free. Once again, if a data transmission contains errors and must be resent, a human being must carry out this action.

cyclic redundancy check (CRC): This error-checking mechanism involves calculating the number of bits in a block of data: it will be equal to some binary number. This number is then divided by some other predetermined binary number. This resulting number is added to the data block at the time of transmission and is checked when the data is received. If the two numbers have the same value, it can be assumed that the transmission is error free. CLC can be used to check for errors in stored data, as well, and can be formatted onto computer storage disks.

ESS See **electronic switching system.**

essential service A network-provided service feature in which a priority dial tone is furnished. Essential service is typically provided to fewer than 10 % of network users.

Ethernet Along with **token ring**, one of the two most prevalent data transmission protocols for LANs. An Ethernet LAN usually operates within an office or building that links computers and peripherals (printers, for example), so resources such as printers, computer terminals, Internet access, and databases can be shared among groups of users. An Ethernet LAN can be constructed either with twisted copper telephone wire or coaxial cable, which has a larger capacity, but is more expensive to install. An Ethernet LAN can operate at up to 10 Mbps; in this format it is referred to as 10BASE-T ("10" signifies 10 Mbps, "BASE" is an abbreviation for **baseband**, and "T" signifies unshielded twisted pair wire). The speed of 10 Mbps for a 10BASE-T Ethernet LAN compares with 16 Mbps for token ring and 100 Mbps for **fiber distributed data interface (FDDI)**. Since an Ethernet LAN is a shared medium—carrying information to and from various users and peripheral devices that are usually less than 300 feet apart—there are occasionally data "collisions" on the network. The Ethernet LAN is arranged to avoid collisions as much as possible because a device attached to the network checks for activity on the network before sending data. The protocol used to sense and ideally avoid collisions is called **carrier sense multiple access with collision detection (CSMA/CD)**. Sometimes, however, two packets are sent at roughly the same time and collide. If this happens and a network node detects it, the collision detection sets in: the first node to recognize the collision sends a signal to all other nodes by jamming the network. After a brief, pre-measured pause, sending nodes try again. Ethernet networks also include a "hub" or "hubs." This is a central wiring point for networks that are shaped like stars—called star-topology networks because all the transmission lines run into this hub, which in an office building is typically located in a closet. In other words, the hub is the central location that connects all the lines that enter it into a single network. Some hubs can be programmed to regenerate or re-time the signals it receives and relays.

Fast Ethernet, also called 100BASE-T10, offers links at speeds of 100 Mbps. Fast Ethernet is usually used for the LAN network backbone, while 10BASE-T is used to connect to individual computers or other data devices. Ethernet is a relatively inexpensive technology for creating networks, and thus is very popular. Ethernet adapter cards for personal computers cost roughly $100. Xerox Corp. developed Ethernet, and registered the name as a trademark. Since then, many other major computer and communications companies have picked up and continued to develop the transmission technology. In order for this LAN technology to become standardized and to be interoperable with alternative LAN technologies, such as token ring and FDDI, it was necessary for an international standards body to become involved. The Institute of Electrical and Electronic Engineers (IEEE) took on the task and developed the "802" committee to evaluate all LAN technologies. This allows both standardization of the individual technologies and creates a method of ensuring all LAN technologies can communicate with one another.

exchange **1.** A room or building equipped so that telephone lines terminating there may be interconnected as required. The equipment may include manual or automatic switching equipment. **2.** In the telephone industry, a geographic area (such as a city and its environs) established by a regulated telephone company for the provision of local telephone services. **3.** In the Modification of Final Judgment (MFJ), a local access and transport area.

extended area service (EAS) A network-provided service feature in which a user pays a higher flat rate to obtain wider geographical coverage without paying per-call charges for calls within the wider area.

extranet A private data network that uses the public telephone network to create a secure Internet-like network among chosen organizations, companies, business partners, suppliers, etc. An extranet extends a company's private network (often referred to as an **intranet**) to outside parties in situations in which both sides can benefit by exchanging information quickly and privately. Like the Internet, an extranet uses the **TCP/IP** protocol and is accessed by users with a **browser** on a PC. Usually, separate organizations' internal intranets are connected through high-speed leased lines secured from the telephone company, thus creating the extranet. Extranets are highly protected with firewalls, encryption, user identification, and other security measures to prevent outsiders from logging onto the network. Various users are given differing degrees of access, based on their identification codes, to corporate information contained in the various workstations and databases connected to the network. Both intranets and extranets typically include Web sites where users can access useful information about their company, job, benefits, suppliers, competitors, etc. Increasingly, businesses are recognizing the benefits of having dedicated links outside their own organization because so-called **electronic commerce** both speeds and reduces the cost of many business transactions. Many major corporations have elected not to use the public Internet as a "free" extranet because of the relative lack of security.

extremely high frequency (EHF) Frequencies from 30 GHz to 300 GHz.

extremely low frequency (ELF) Frequencies from 30 Hz to 300 Hz.

F

facilities-based carrier A telephone company that owns at least some of its own network, switching, and transmission facilities. The facilities it doesn't own it may lease from other carriers as required. Contrast with a **switchless reseller**, which owns no facilities and simply resells phone service purchased wholesale from facilities-based carriers.

facsimile (FAX) The process by which fixed graphic images, such as printed text and pictures, are scanned, then converted into electrical signals that may be transmitted over a telecommunications system and used to create a copy of the original, or an image so produced. Wirephoto and telephoto are facsimile via wire circuits. Radiophoto is facsimile via radio. Technology now exists that permits the transmission and reception of facsimile data to or from a computer without requiring a hard copy at either end. Current facsimile systems are designated and defined as follows:

Group 1 Facsimile: The mode of black and white facsimile operation, defined in CCITT Recommendation T.2, that uses double sideband modulation without any special measures to compress the bandwidth. An 8½×11-inch document may be transmitted in approximately 6 minutes via a telephone-type circuit. Additional modes in this group may be designed to operate at a lower resolution suitable for the transmission of documents in 3 to 6 minutes.

Group 2 Facsimile: The mode of black and white facsimile operation, defined in CCITT Recommendation T.3, that accomplishes bandwidth compression by using encoding and vestigial sideband, but excludes processing of the document signal to reduce redundancy. An 8½ × 11-inch document may be transmitted in approximately 3 minutes using a 2100-Hz AM/PM/VSB, over a telephone-type circuit.

Group 3 Facsimile: The mode of black and white facsimile operation, defined in ITU-T Recommendation T.4, that incorporates means for reducing the redundant information in the signal by using a one-dimensional run-length coding scheme prior to the modulation process. An 8½ × 11-inch document may be transmitted in approximately 1 minute or less over a telephone-type circuit with twice the Group 2 horizontal resolution. The vertical resolution may also be doubled. Group 3 Facsimile machines have integral digital modems.

Group 3C Facsimile: The Group 3 digital mode of facsimile operation defined in CCITT Recommendation T.30. Group 3C is also referred to as Group 3 Option C or as Group 3-64 Kbps.

Group 4 Facsimile: The mode of black and white facsimile operation defined in ITU-T Recommendation T.563 and CCITT Recommendation T.6. Group 4 Facsimile uses bandwidth compression techniques to transmit, essentially without errors, an 8½ × 11-inch document at a nominal resolution of 8 lines/mm in less than 1 minute over a public data network voice-grade circuit. When any CCITT or CCIR Recommendation is modified by the ITU-T, the modified document is designated as an ITU-T Recommendation.

Fast Ethernet See **Ethernet**.

fast-packet switching A packet switching technique that increases the throughput by eliminating overhead. Overhead reduction is accomplished by allocating flow control and error correction functions to either the user applications or the network nodes that interface with the user. Cell relay and frame relay are two implementations of fast-packet switching.

FAX See **facsimile**.

fax back See **fax-on-demand**.

fax broadcasting An automated system for broadcasting the same fax message to a large number of recipients.

fax mailbox A telephone company service that operates much like voice mail, except faxes, instead of voice recordings, are stored for later retrieval. A subscriber to this service can call into the designated service number, and using the touch-tone keypad, enter his or her PIN and the phone number where he or she wants the fax sent. This service is intended for people on the move who need to be able to forward their fax messages to wherever they happen to be.

fax modem A modem that performs the same function as a stand-alone fax machine. It can be internal to the computer, a PCMCIA card for a laptop computer, or a separate external box attached to a PC. It is ideally suited for sending outgoing faxes because it eliminates the need to print the document before running it through a regular fax machine, and the image is usually superior because it is created directly by the computer without the intervening steps required with a regular fax machine. Incoming faxes are received as images and must often be printed to be legible. Many people use a fax modem to send faxes while relying on their regular fax machine for incoming faxes.

fax-on-demand An automated, computer-based system that allows a caller to select documents, usually from a menu of selections using the touch-tone keypad, to be transmitted back to his or her fax machine. This service can be offered in two variations: 1) The caller can dial into the system from any touch-tone phone, make the document selection(s), and designate the fax number where the fax is to be sent, or 2) The caller must call from the handset on the fax machine where the fax is to be received, and the fax will be transmitted when the caller is prompted to press the "start" key. The second version has the advantage to the fax service provider of not requiring an outgoing toll call to send the fax, so that the caller pays for the entire telephone call, unless, of course, the service is offered on a toll-free 800 number.

fax server A computer with a large hard drive that stores incoming or previously loaded fax document images for retrieval by addressees, subscribers, or others with authorized access to the server. Fax-on-demand systems and fax mailbox services use fax servers.

FCC See **Federal Communications Commission.**

FCC registration program The Federal Communications Commission program and associated directives intended to assure that all connected terminal equipment and protective circuitry will not harm the public switched telephone network or certain private line services. The FCC registration program requires the registering of terminal equipment and protective circuitry in accordance with Subpart C of part 68, Title 47 of the *Code of Federal Regulations.* This includes the assignment of identification numbers to the equipment and the testing of the equipment. The FCC registration program contains no requirement that accepted terminal equipment be compatible with, or function with, the network.

FDDI See **fiber distributed data interface.**

FDMA See **frequency-division multiple access.**

Federal Communications Commission (FCC) The U.S. Government board of five presidential appointees that has the authority to regulate all non-federal government interstate telecommunications (including radio and television broadcasting) as well as all international communications that originate or terminate in the United States.

fiber distributed data interface (FDDI) A LAN technology that uses fiber-optic lines for transmission, rather than the coaxial cable or copper wire used on **Ethernet** LANs. Fiber optic is the transmission medium and technology that allows information to be transmitted as light along a strand of glass or fiber or a plastic wire. Light transmissions can move much more data faster than alternate communication technologies, and telephone companies, particularly long-distance carriers, are rapidly converting their networks to fiber. FDDI runs at 100 Mbps using a **token ring** technology, and is often used as the LAN backbone—connecting other LANs together at distances up to 124 miles. (In terms of speed, an FDDI LAN is ten times faster than an Ethernet LAN.) FDDI networks actually contain two token rings; one can be used for backup in the event of a failure on the primary ring. If the network is functioning properly, the second ring can be used to increase the transmission capacity of the FDDI network. See also **token ring**.

fiber-optic cable A telecommunications cable in which one or more optical fibers are used as the propagation medium. A fiber-optic cable may be an all-fiber cable, or contain both optical fibers and metallic conductors. One possible use for the metallic conductors is the transmission of electric power for repeaters. Strands of glass filaments transmit data using flashes of laser light. This is a relatively low-cost means of transmitting large amounts of data for long distances with little distortion. The backbones of most long-distance networks are built with fiber-optic cable. The first generation of commercial long-distance fiber-optic networks appeared in the 1980s. The technology has advanced since then. Older cables carried flashes of light at a single wavelength, or color. The newest cables have eight windows, or streams of light, operating with different colors. Each color stream within the fiber carries about 10 gigabits of information. The division of the cable into multiple lanes is achieved by a compression technique known as *wavelength-division multiplexing*. It is still evolving. The next generation of fiber will have 16 windows, each capable of carrying 40 gigabits of information a second.

file server A specialized computer with a large hard drive where all the organization's files and application software reside. It is connected to all the PCs in the organization via a LAN. With one central repository for all data and programs, it is easier to backup data periodically, to install software upgrades (only once, on the server), and for multiple users to access the most up-to-date files. Further, fewer peripheral devices such as modems and printers are needed because they can be connected to the server and be shared by everyone.

File Transfer Protocol (FTP) The Transmission Control Protocol/Internet Protocol (TCP/IP) protocol that is (a) a standard high-level protocol for transferring files from one computer to another, (b) usually implemented as an application level program, and (c) uses the Telnet and TCP protocols. In conjunction with the proper local software, FTP allows computers connected to the Internet to exchange files, regardless of the computer platform.

firewall A secure gateway between networks, often between a private network and the Internet. The National Institute of Standards and Technology (NIST) describes a firewall as a set of related programs, all of which are located at a gateway server to a network, that is used to protect a private network from access by users from other networks. For example, most corporate intranets that allow employees to use the public Internet through their private network contain firewalls both to control the resources employees can access on the Internet as well as to prevent intruders from gaining entry to the corporate intranet. Firewalls work with a **router** program to filter network packets, evaluate them and determine whether they meet the criteria to be sent on to their destination. The firewall is often installed in a separate computer, so that all incoming requests must come through this separate device rather than getting directly into private network resources. Elements in a firewall used for security include screening domain names and IP addresses, as well as coordinating secure log-on procedures. Firewalls also include reporting functions, alarms that indicate the gateway may have been compromised, and a specially designed computer screen (graphical user interface) used to control the firewall. The NIST is careful to explain that a firewall is more than a router or collection of systems providing security to a network. "A firewall is an approach to security, it helps implement a larger security policy that defines the services and access to be permitted, and it is an implementation of that policy in terms of network configuration, one or more host systems and routers, and other security measures such as advanced authentication in place of static passwords. The main purpose of a firewall system is to control access to or from a protected network. It implements a network access policy by forcing connections to pass through the firewall, where they can be examined and evaluated," according to NIST. Firewalls can be placed at the network level to examine entering and exiting traffic at the link between two networks, or at the application level of the network to control access to applications on the LAN or Internet (such as e-mail).

five hundred (500) service A telephone service that allows individuals to receive, via a single number, telephone calls in various locations (*e.g.*, home, office, or car phone) from call originators not necessarily using the same common carrier.

fixed wireless A transmission system where radio signals are employed in the local loop, taking the place of the copper wires that comprise the traditional local loop service from the local exchange carrier (LEC). The subscriber has a receiving antenna attached to its premises, where radio signals transmitted from the service provider are converted back into electrical signals and pass into the existing telephone wiring inside the building. From the subscriber's point of view the service is no different from regular phone service. This is in contrast to *mobile* wireless systems such as analog cellular or PCS systems. Service providers employ fixed wireless systems to avoid building a wired infrastructure or purchasing network elements from an established LEC. Also known as **wireless local loop** (WLL). See also **local multipoint distribution service** (LMDS).

flat rate service Telephone service in which a single payment permits an unlimited number of local calls to be made without further charge for a specified period of time.

FM See **frequency modulation**.

follow-me services This service is usually but not always offered using a 500 number service access code. It can be programmed by the subscriber to dial a series of preprogrammed numbers where the subscriber is likely to be, and then default to voice mail or a pager number as the final option. The caller need dial only one number (a 500 number, for example), and the service will dial the programmed numbers in succession, such as office, home, cell phone, and then voice mail. A recording will usually let the caller know the status of the search with a message such as, "Please stand by, we are trying your party at another location."

foreign exchange (FX) service A network-provided service in which a telephone in a given local exchange area is connected, via a private line, to a central office in another, i.e., "foreign" exchange, rather than the local exchange area's central office. To callers, it appears that the subscriber having the FX service is located in their local exchange area. This is useful for companies that want a perceived presence in a certain area, with a listing in the local telephone directory.

forward channel The channel of a data circuit that transmits data from the originating user to the destination user. The forward channel carries message traffic and some control information.

fractional T1 A data rate corresponding to telephone carrier leased-line services offering a portion of the 24 channels available on a full T-1 line, typically between 56 Kbps (DS0) and 1.554 Mbps. Typically provided on two pairs of copper wires, Fractional T1 rates are less expensive than full T1 and appeals to users wishing to connect remote LANs, for example, or to support videoconferencing services on their networks. See also **T1**.

frame Usually a series of data bits sent over a communications line containing information (the essential data being transmitted between users), plus beginning and ending marker bits (called *flags*), and some combination of bits containing address or error checking information, a control field, or frame-check sequence.

frame relay A fast-packet data communications standard that allows a network to carry data frames in packets of varying length; usually used to connect LANs or for LAN-to-WAN connections. Frame relay networks are high-speed networks that can link with and carry traffic delivered by networks using a variety of other fast-packet technologies, including X.25, ATM, and B-ISDN. It is similar to X.25 networks in that it employs packet switching technology. Unlike X.25 networks, frame relay networks have the ability to carry packets of many different lengths. This distinction is important because a frame relay network can therefore accommodate data packets associated with virtually any data protocol, even though they vary in size. For example, an X.25 packet ranges in size from 128 bytes to 256 bytes, while an Ethernet frame can be 1,500 bytes—a frame relay network is able to carry either.

Frame relay networks are typically faster and less expensive to operate than X.25 networks for a variety of reasons. First, unlike X.25 networks, a frame relay network does not attempt to convert the data protocols of various transmissions to suit an assigned data protocol. Frame relay networks are called *protocol independent* because they can accept any type of data, then switch and carry it on the network. Any protocol conversion of data is the responsibility of the users transmitting and receiving information on a frame relay network. An X.25 network, on the other hand, converts data packets it transmits to a fixed protocol. Again in contrast to X.25, which checks data for errors throughout its system of network nodes, data sent on a frame relay network is not checked for errors or lost packets, and any errors must be detected and corrected by the user.

Frame relay's ability to accommodate virtually any data protocol without investing the time or computer energy to convert it to a particular protocol makes it a less expensive, faster network. In a similar fashion, pushing error detection/correction back to the user also results in faster, less expensive transmission. Connections in a frame relay network are described as *permanent virtual connections*—virtual because they are not dedicated to particular users. Instead, a frame relay network is shared among users and is capable of creating an exclusive circuit through this shared network. A shared network simply means that switches on the network accept data packets as they come, read the addresses and forward the data. X.25 networks also function this way. A frame relay network is more vulnerable to delays (called *latency*) in delivering data because the length of each packet can vary so considerably. Because packets on an X.25 network are fixed in length, these kinds of delays are somewhat less likely to occur. Frame relay is most often used by large, corporate customers to interconnect LANs or to connect a LAN to a WAN. Frame relay is especially suited to data traffic that is described as *bursty,* meaning packets are sent in uneven intervals.

frequency-division multiple access (FDMA) A technology allowing more than one signal to be sent on a communications channel, such as a satellite transponder, simultaneously. Each user's transmission is assigned a dedicated frequency that will not interfere with frequencies of other users. FDMA is accomplished using **frequency-division multiplexing**.

frequency-division multiplexing (FDM) A rather complex technology allowing separate signals to share a transmission line by dividing the signals into different frequencies and then recombining them for transmission in a process called **multiplexing**. These combined signals can be sent on the network as a single signal, creating appealing network economies. A device at the receiver end separates the signals that were multiplexed together. Radio, television, and cable service use frequency-division multiplexing, and FDM was commonly used in the older analog long-distance networks to allow each circuit to carry more conversations. As long-distance providers have added fiber optics, however, FDM is being replaced by a newer process called **time-division multiplexing (TDM)**.

frequency modulation (FM) Modulation in which the instantaneous frequency of a sine wave carrier is caused to depart from the center frequency by an amount proportional to the instantaneous value of the modulating signal. In FM, the carrier frequency is called the center frequency. FM is a form of angle modulation. In optical communications, even if the electrical baseband signal is used to frequency-modulate an electrical carrier (an "FM" optical communications system), it is still the intensity of the lightwave that is varied (modulated) by the electrical FM carrier. In this case, the "information,"as far as the light wave is concerned, is the electrical FM carrier. The light wave is varied in intensity at an instantaneous rate corresponding to the instantaneous frequency of the electrical FM carrier.

frequency range A continuous range or spectrum of frequencies that extends from one limiting frequency to another. The frequency range for given equipment specifies the frequencies at which the equipment is operable. For example, filters pass or stop certain bands of frequencies. The frequency range for propagation indicates the frequencies at which electromagnetic wave propagation in certain modes or paths is possible over given distances. Frequency allocation, however, is made in terms of bands of frequencies. There is little, if any, conceptual difference between a range of frequencies and a band of frequencies.

frequency-shift keying (FSK) Frequency modulation for digital signals in which the modulating signal shifts the output frequency between two predetermined values. Unlike frequency modulation for analog signals, in which the frequency varies continuously over a range, with FSK the frequency is shifted between only two discrete values, termed the *mark* and *space* frequencies, representing the binary values of 1 or 0. Also called *frequency-shift modulation, frequency-shift signaling*.

frequency-shift modulation See **frequency-shift keying.**

FTP See **File Transfer Protocol.**

full-duplex (FDX) circuit A circuit that permits simultaneous transmission in both directions, *i.e.*, moving in opposite directions at the same time. Two-party phone calls are best made on four-wire communications lines, which are FDX circuits. This allocates two wires for communications in one direction and two wires for communication in the opposite direction. Long-distance lines are four wires. In local telephone communications, however, most lines are two wires, allocating just one wire for transmissions from point B to point A and vice versa. This results in lower-quality communications. A speakerphone is often *half-duplex,* which means at any given time they can transmit in just one direction—away from the person speaking loudest. FDX speakerphones are available.

G

gatekeeper A function built into network elements such as gateways and virtual PBXs that controls access to the data network, denying access to users when the traffic is too heavy to accept additional packets. The user essentially gets a busy signal.

gateway In a communications network, a specialized network node equipped for interfacing with another network that uses different protocols, providing translation functions necessary for interoperability between dissimilar networks. Gateways, which can be specialized computers, can contain a variety of other communications devices, such as rate converters or signal translators to facilitate communication between the two networks. E-mail messages often pass through various gateways as they move from a LAN to the public Internet and perhaps on to a specialized e-mail system like America Online.

geostationary satellite A geosynchronous satellite whose circular and direct orbit lies in the plane of the Earth's equator and which thus remains fixed relative to the Earth; by extension, a satellite that remains approximately fixed relative to the Earth.

gigahertz (GHz) A unit of frequency denoting 10^9 Hz.

Global Standard for Mobile communications (GSM) The original term is French: Groupe Speciale Mobile (GSM). The most widely used of three digital wireless telephone technologies. [The other two are **Time Division Multiple Access** (TDMA) and Code Division Multiple Access (CDMA)]. GSM has been chosen as the standard for digital cellular telephony in scores of countries, including Europe and Japan. While not a standard in the United States, GSM is becoming more popular in the U.S. It is primarily used for broadband personal communication services (PCS) in the U.S.

GSM is based on TDMA, which is a method of mobile communication that allows a large number of users to access a single radio frequency channel. To prevent the interference that would occur if the transmissions overlapped, each user is assigned a unique time slot. This ensures the digital transmissions are sent in sequence. Like TDMA, GSM equipment compresses digitized data before transmitting it over the assigned radio frequency channel. Under GSM, a communications channel is shared with two other users.

GSM subscribers carry a *smart card,* also called a *subscriber identity module (SIM)* card that contains their user and account information. Customers can use public GSM telephones by inserting their SIM card to activate it. The telephone is able to "read" the card and identify the user to the network, ensuring delivery of appropriate service and accurate bills. This smart card system is convenient for users, who do not have to carry their own telephone since they can use any public GSM phone.

Organizations in different countries have agreed to ensure their GSM networks remain interoperable so a mobile user in Europe, for example, could continue a call when crossing the national boarder into another country.

Government Emergency Telecommunications Service (GETS) This is an emergency, toll-free government service that uses the 710 numbering plan area (NPA) code. It was designed for use during times of emergencies such as natural disasters when telephone network congestion or outages occur. Only authorized users (i.e., emergency relief workers) with PIN codes can access the service, which allows them to conduct urgent communications with enhanced routing and priority treatment within the telephone network.

half-duplex (HDX) operation Operation in which communication between two terminals occurs in either direction, but in only one direction at a time. Half-duplex operation may occur on a half-duplex circuit or on a duplex circuit, but it may not occur on a simplex circuit.

handoff **1.** In cellular mobile systems, the process of transferring a phone call in progress from one cell transmitter and receiver and frequency pair to another cell transmitter and receiver using a different frequency pair without interruption of the call. **2.** In satellite communications, the process of transferring ground-station control responsibility from one ground station to another without loss or interruption of service.

handshaking **1.** In data communications, a sequence of events governed by hardware or software, requiring mutual agreement of the state of the operational modes prior to information exchange. **2.** The process used to establish communications parameters between two stations. Handshaking follows the establishment of a circuit between the stations and precedes information transfer. It is used to agree upon such parameters as information transfer rate, alphabet, parity, interrupt procedure, and other protocol features.

HDSL See **high speed digital subscriber line**.

head end **1.** A central control device required by some networks (*e.g.,* LANs or MANs) to provide such centralized functions as remodulation, retiming, message accountability, contention control, diagnostic control, and access to a gateway. **2.** A central control device, within CATV systems, that provides centralized functions such as remodulation.

hertz (Hz) A unit of frequency equal to one cycle per second. A periodic phenomenon that has a period of one second has a frequency of one hertz.

HF See **high frequency**.

HFC See **hybrid fiber coax**.

hierarchical routing Routing that is based on hierarchical addressing. Most Transmission Control Protocol/Internet Protocol (TCP/IP) routing is based on a two-level hierarchical routing in which an IP address is divided into a network portion and a host portion. Gateways use only the network portion until an IP datagram reaches a gateway that can deliver it directly. Additional levels of hierarchical routing are introduced by the addition of subnetworks.

high frequency (HF) Frequencies from 3 MHz to 30 MHz.

high speed digital subscriber line (HDSL) A wideband data service that uses twisted-pair copper telephone lines to link telephone companies to corporate networks as well as to link sites within a corporate network. Major corporations are using HDSL because it allows a single twisted-pair to operate at speeds comparable to T1 (1.536 Mbps), which is more expensive and more difficult to deploy. HDSL is a symmetric service (in contrast to **ADSL**), providing equal bandwidth in each direction. As a result, HDSL is a slower service (less bandwidth) than ADSL, which can carry data at theoretical speeds up to 8.5 Mbps. See also **digital subscriber line**.

hold call A service feature that lets you place a call on hold even if you don't have a hold button on your phone. You do this by pressing **hookflash**, dialing the call hold code and hanging up. The call is placed on hold and is resumed when you pick up the handset at the original phone or at an extension.

hookflash This is performed by depressing and releasing the **hookswitch** button or by pressing a "flash" button on the telephone set. Some phone company services require a hookflash to activate certain features.

hookswitch The button that is depressed when you place the phone receiver in its cradle, terminating the connection/call. Also called switchhook.

HTML See **Hypertext Markup Language**.

HTTP See **Hypertext Transfer Protocol**.

hub A central distribution point on a network from which data can be sent and received via one or more communication lines connected to it. The name "hub" is derived from a wheel, which has a central hub with some number of spokes radiating out from it. A switch is usually incorporated into a hub so communications coming into the hub can be forwarded (switched) in a variety of directions. The term *hub* is also used interchangeably with *switch*, in some instances, and *router* in others.

hunt group In telephony, a group of lines that are accessed sequentially to find and establish a connection with an idle circuit of a chosen group.

hybrid cable An optical communications cable having two or more different types of optical fibers, *e.g.,* single-mode and multimode fibers.

hybrid fiber coax (HFC) A cabling scheme offering high bandwidth capacity that employs a combination of fiber optic and coaxial cable. The fiber-optic cable provides the long-haul backbone while the coaxial cable branches off to serve individual subscribers.

hybrid phone system This phone system combines features of both key and PBX systems. Unlike key systems, you can install nonproprietary, regular phones on some extensions or stations, and every line does not have to be available at every phone. Like PBX systems, some hybrids will accommodate an automated attendant to route calls.

Hypertext Markup Language (HTML) An application of SGML (Standard Generalized Markup Language [ISO 8879]) implemented in conjunction with the World Wide Web to facilitate the electronic exchange and display of simple documents using the Internet.

Hypertext Transfer Protocol (HTTP) In the World Wide Web, a protocol that facilitates the transfer of hypertext-based files between local and remote systems.

Hz See **hertz.**

I

ICI See **incoming call identification.**

ICMP See **Internet Control Message Protocol.**

IDN See **integrated digital network.**

IDSL ISDN-digital subscriber line. One of the xDSL offerings that uses ISDN technology to deliver dedicated data services, as opposed to the standard circuit-switched ISDN service that delivers digital voice, data, and video. See **digital subscriber line**.

IN See **intelligent network.**

incoming call identification (ICI) A switching system feature that allows an attendant to identify visually the type of service or trunk group associated with a call directed to the attendant's position.

incumbent local exchange carrier (ILEC) This term was created as a result of the Telecommunications Act of 1996 to differentiate between the *established* local phone company and a new entrant into the market, known as a **competitive local exchange carrier (CLEC)**.

individual line A line that connects a single user to a switching center.

information provider (IP) A business or entity that delivers information or entertainment services to end users (callers), usually over 800 or 900 lines, with the use of communications equipment and computer facilities. The IP charges the callers at a rate above the underlying transport charges paid to the telephone company in order to cover program costs and to provide a profit.

inside plant All the cabling, switches, and equipment installed in a telecommunications facility, including the main distribution frame (MDF) and all the equipment extending inward therefrom, such as PBX or central office equipment, MDF protectors, and grounding systems.

integrated digital network (IDN) A network that uses both digital transmission and digital switching.

integrated services digital network (ISDN) A set of communications protocols that converts one analog telephone line (one pair of copper wires) into multiple digital communication channels with much more capacity. ISDN allows digital data—audio, video, or text—to be delivered over the copper-wire telephone network. Since a single ISDN line has at least three channels, ISDN customers can use the same line to perform different tasks simultaneously. For example, users with ISDN service can go online, talk on the telephone, and receive a fax simultaneously. Telephone companies are pursuing ISDN as one of several technologies that support digital service on the existing public telephone network, a predominantly analog network built using copper wire.

Phone companies and network equipment manufacturers have been working on ISDN since the early 1980s. ISDN is available in a variety of configurations. The two most widely deployed versions of ISDN are **Basic Rate Interface** (BRI-ISDN), for residential and small-business customers, and **Primary Rate Interface** (PRI-ISDN), for users who need more capacity.

Currently, digital transmissions must be translated into analog signals for transport over the telephone network. For example, when you send an e-mail message from your computer, your modem modulates the message, meaning the modem changes the information your computer has generated from a digital format into an analog format. When the e-mail message arrives at its destination, typically another computer connected to the telephone network in another location, the modem on the recipient's computer demodulates the transmission, turning it back into a digital format that a computer can read. Modems were an exciting breakthrough, allowing remote computers to "talk" to each other over the global telephone network. But analog telephone lines are limited in capacity. People using the newest 56 Kbps modems already are bumping up against the innate limitation of using an analog transport mechanism to deliver digital information.

An alternative, which many phone companies are pursuing, is for phone companies to create all-digital networks by replacing copper wire with fiber-optic lines and upgrading analog network equipment to digital gear. But this is an expensive and time-consuming endeavor.

In a sense, ISDN is a bridging technology because it allows legacy (existing) analog phone lines to support digital communication services and applications. Unlike modem technology, which changes the information being delivered to the appropriate format (digital for computer, analog for the phone network), ISDN technology converts portions of the network itself from analog to digital. A digital network can support a host of new communication services that cannot be deployed on an analog network, such as affordable desktop videoconferencing, collaborative computing, call waiting, calling line identification, and others. But if ISDN is such a beneficial service, why has it been slow to gain widespread acceptance?

First, the local telephone companies themselves were slow to make the network upgrades required to support the service, which was introduced in the early 1980s. Service providers did not know how strong demand for ISDN would be and the necessary hardware and software was expensive; this made them reluctant to invest in the upgrade. In recent years phone companies have been deploying ISDN more quickly and it is available in most urban areas and major suburbs. There are still rural areas, however, that do not have the service.

Second, assuming ISDN is locally available to a would-be customer, setting up the service for a home or small business can be a daunting task, often taking a month or more and requiring expert assistance. Until a few years ago, there were no specifications for standard ISDN service or equipment. As a result, ISDN equipment made by different manufacturers and ISDN service offered by different phone companies rarely were compatible. Customers had to purchase ISDN equipment that was designed specifically for use on their local exchange carrier's network. There still is not an official ISDN standard, but there are specifications for standard equipment and service. Today, virtually all ISDN devices and services are compatible—in the U.S. The standards for ISDN in other parts of the world are still different.

Third, even though costs have come down, the equipment customers must buy to use ISDN is expensive. An ISDN network termination device, which users must have, costs about $250. The network termination device, also called NT1, acts as an interface between the phone company's ISDN line and the user's ISDN phone, ISDN modem, or ISDN adapter. Unlike regular telephone service, which allows us to simply plug a telephone or modem into the jack on the wall, ISDN users must plug their computers or telephones into an NT1, which in turn plugs into the service.

An ISDN terminal adapter is different than an NT1. The terminal adapter is a device that allows analog phones or fax machines to communicate over an ISDN line. A specialized terminal adapter, called an *ISDN modem*, allows analog devices to communicate over an ISDN line and also allows ISDN devices to communicate with analog modems.

Some telecommunications analysts believe that ISDN will ultimately become an interim technology used in the public network to boost performance until other digital services can be deployed. The two digital public-network technologies most widely considered likely successors to the POTS network (plain old telephone service) are synchronous optical network (SONET) and advanced intelligent network (AIN). ISDN, however, has gained momentum as a service to support high-speed Internet access and broadband for the home office and small business.

integrated station A terminal device in which a telephone and one or more other devices, such as a video display unit, keyboard, or printer, are integrated and used over a single circuit.

integrated system A telecommunication system that transfers analog and digital traffic over the same switched network.

integrated voice and data terminal (IVDT) See **integrated station.**

intelligent network (IN) **1.** A network that allows functionality to be distributed flexibly at a variety of nodes on and off the network and allows the architecture to be modified to control the services. **2.** In North America, an advanced network concept that is envisioned to offer such things as (a) distributed call-processing capabilities across multiple network modules, (b) real-time authorization code verification, (c) one-number services, and (d) flexible private network services including reconfiguration by subscriber, traffic analyses, service restrictions, routing control, and data on call histories. Levels of IN development are identified below:

IN/1: A proposed intelligent network targeted toward services that allow increased customer control and that can be provided by centralized switching vehicles serving a large customer base.

IN/1+: A proposed intelligent network targeted toward services that can be provided by centralized switching vehicles, *e.g.,* access tandems, serving a large customer base.

IN/2: A proposed, advanced intelligent-network concept that extends the distributed IN/1 architecture to accommodate the concept called *service independence*. Traditionally, service logic has been localized at individual switching systems. The IN/2 architecture provides flexibility in the placement of service logic, requiring the use of advanced techniques to manage the distribution of both network data and service logic across multiple IN/2 modules.

interactive In audiotext, a capability that allows the caller to select options from a menu of programmed choices in order to control the flow of information. As the term implies, the caller truly interacts with the computer, following the program instructions and selecting the information he or she wishes to receive. See also **interactive voice response**.

interactive voice response (IVR) The telephone keypad substitutes for the computer keyboard, allowing anyone with a touch-tone telephone to interact with a computer. Instead of displaying information on a computer screen, IVR uses a digitized voice to convey the desired information to the caller, who follows the voice prompts and menu selections to get the desired results.

intercept **1.** To stop a telephone call directed to an improper, disconnected, or restricted telephone number, and to redirect that call to an operator or a recording. **2.** To gain possession of communications intended for others without their consent, and, ordinarily, without delaying or preventing the transmission. An intercept may be an authorized or unauthorized action. **3.** The acquisition of a transmitted signal with the intent of delaying or eliminating receipt of that signal by the intended destination user.

intercom A telephone apparatus by means of which personnel can talk to each other within a building or facility. An intercom can be a stand-alone system or incorporated into a telephone system. When part of a phone system, it usually features the capability of dialing all stations connected to the system with abbreviated dialing of one to four numbers.

interconnect agreement An agreement between telephone companies (usually the incumbent local exchange carrier [ILEC] and a competitive local exchange carrier [CLEC]) that allows the subscribers of both companies to access their respective networks and call one another.

interconnect (company) A company that sells and installs telephone systems (*i.e.*, sells PBXs to businesses) and "interconnects" the phone systems to the public network.

interconnect facility In a communications network, one or more communications links that (a) are used to provide local area communications service among several locations and (b) collectively form a node in the network. An interconnect facility may include network control and administrative circuits as well as the primary traffic circuits.

interconnection **1.** The linking together of interoperable systems. **2.** The linkage used to join two or more communications units, such as systems, networks, links, nodes, equipment, circuits, and devices.

interexchange carrier (IXC) A communications **common carrier** that provides telecommunications services between LATAs or between exchanges within the same LATA. Interexchange carriers have usually relied on local exchange carriers or competitive access providers for the local origination and termination of their traffic. Basically, long-distance phone companies such as AT&T, MCI and Sprint.

inter-LATA Between local access and transport areas (LATAs). Services, revenues, and functions associated with telecommunications that originate in one LATA and that terminate in another one or that terminate outside of that LATA.

interleaving The transmission of pulses from two or more digital sources in time-division sequence over a single path. Picture two (or more) highway on-ramps where cars are allowed onto the main highway in an alternating fashion.

intermediate distribution frame (IDF) In a central office or customer premises, a frame that (a) cross-connects the user cable media to individual user line circuits and (b) may serve as a distribution point for multipair cables from the main distribution frame (MDF) or combined distribution frame (CDF) to individual cables connected to equipment in areas remote from these frames. See also **distribution frame**.

internal call A call placed within a private branch exchange (PBX) or local switchboard, *i.e.*, not through a central office in a public switched network.

international callback Used by businesses and travelers in foreign countries that have high international calling rates. Using a pre-assigned, dedicated phone number, you dial your callback service, located in the United States, and hang up after one ring. Because you hang up after only one ring, without actually completing the call, the local phone company cannot charge you for a completed call. However, that one ring is enough to signal the switch/computer that you called, seeking dial tone to set up a call.

Recognizing your preassigned call-in number, the switch/computer is programmed to immediately dial you back at the phone number you previously designated, referred to as your *callback number*. You now have dial tone that originates in the U.S. Then you will be prompted by the switch/computer to enter the phone number you want to call. Once this is accomplished, the call is set up, consisting of two legs that both originate in the U.S: one to your location and another to the party you are calling.

international simple resale (ISR) The use of leased or resold international private lines, which are interconnected with the public switched telephone network at each end, in providing international switched long-distance service. This can save a lot of money because it avoids the system of **accounting rates** and **settlement rates** that drive up the cost of international calling. The FCC permits ISR between the U.S. and several foreign countries (based on meeting certain tests regarding settlement rates), which combined account for nearly 50% of the traffic from the U.S.

International Organization for Standardization (ISO) An international organization that (a) consists of member bodies that are the national standards bodies of most of the countries of the world, (b) is responsible for the development and publication of international standards in various technical fields, after developing a suitable consensus, (c) is affiliated with the United Nations, and (d) has its headquarters at 1, rue de Varembé, Geneva, Switzerland. Member bodies of ISO include, among others, the American National Standards Institute (ANSI), the Association Française de Normalisation (AFNOR), the British Standards Institution (BSI), and the Deutsche Institut für Normung (DIN).

International Telecommunication Union (ITU) A civil international organization established to promote standardized telecommunications on a worldwide basis. The ITU-R and ITU-T are committees under the ITU. The ITU headquarters is located in Geneva, Switzerland. While older than the United Nations, it is recognized by the U.N. as the specialized agency for telecommunications.

International Telegraph and Telephone Consultative Committee See **CCITT, ITU-T.** A predecessor organization of the ITU-T.

internet An interconnection of networks.

[The] Internet A worldwide interconnection of individual networks operated by government, industry, academia, and private parties. The Internet originally served to interconnect laboratories engaged in government research, and has now been expanded to serve millions of users and a multitude of purposes.

Internet Control Message Protocol (ICMP) An Internet protocol that reports datagram delivery errors. ICMP is a key part of the TCP/IP protocol suite. The packet internet gopher (ping) application is based on ICMP.

Internet Protocol (IP) The method (or protocol) used to route information sent from one computer to another on the Internet or other data networks, such as corporate intranets or industry extranets. This is currently a "best effort" protocol with no quality-of-service (QoS) guarantees about how quickly or reliably the data will reach its destination. The IP is a "connectionless" protocol, meaning there is no set connection between a sending and receiving point on the network. Just as there is more than one way for you to drive to work, there are different ways for information to travel from Point A to Point B on the Internet.

The IP oversees a relay race, of sorts, with data packets. When you send an e-mail message on the Internet, the address for the recipient's computer is unique among the millions of computers linked to the Internet. Your message is broken into small pieces called *packets*. Each packet contains your address and the recipient's address. Each packet is sent to a **gateway** computer (not necessarily the same gateway computer) that functions like a neighborhood traffic cop. The gateway computer is familiar with all the addresses in its neighborhood, or domain, as well as the gateways that are adjacent to it. When a gateway receives a packet from your e-mail message, it reads the recipient's address. If the gateway determines that the address is not in its domain, or neighborhood, it relays the message to an adjacent gateway, where the same procedure is followed. Each packet from your message could follow a different route through the Internet because the gateways consider each packet an independent unit.

The gateways are not concerned with delivering your message in its entirety, but only with the individual packets it receives. Ultimately, the packets reach a gateway monitoring the domain in which your recipient's computer resides. This gateway forwards the packets it receives directly to the address of the recipient's computer. Because the individual packets of your message could have followed different routes—some faster or slower, shorter or longer—they are likely to arrive at their final destination in an order that differs from the original message. That is not the Internet protocol's responsibility—its job ends when all the packets arrive at their destination.

The "TCP" part of the familiar TCP/IP protocol, which stands for Transmission Control Protocol, is the method used to reassemble the packets into their original order.

Internet service provider (ISP) An organization that provides access to the Internet. Both individuals and corporations subscribe to ISPs. Most ISPs provide additional services, including giving each subscriber space on their server (computer) for an electronic mail box, access to news groups, and maintaining a Web page that subscribers can select as their browser's home page. The home page typically contains local information, like weather reports and restaurant reviews. ISPs charge by the month for unlimited access or by the hour for actual time subscribers are connected to the service. Independent businesses, on-line service providers, such as AOL, and phone companies offer this service.

Internet telephony Using the Internet, rather than the public switched telephone network (PSTN) for voice or fax communication. Internet telephony appeals to some people because long-distance or international phone calls can be placed over the Internet for the price of the local phone call (placed to access the Internet). A personal computer can function as a telephone with the addition of a telephony board and software that links the PC with phone systems. Internet telephony can also refer to a new Internet marketing/customer service tool. Some companies are adding "call me" buttons to their Web sites. When a visitor clicks on the button, they are requesting a call from a sales or customer service representative. The call is usually a PSTN call. See also **IP telephony**.

Internet telephony service provider (ITSP) A company that offers voice telephone services over the Internet or private IP (Internet protocol) data networks. This service is usually offered in two forms: 1) computer to telephone, and 2) telephone to telephone. The telephone is any regular analog phone connected to the public switched telephone network (PSTN). In the telephone-to-telephone application, for example, the caller dials an access number to connect to the nearest ITSP gateway, which takes the traffic off the PSTN, converts the signals to IP (Internet protocol) data packets, and routes it over the Internet or a private IP network to another gateway nearest the destination, where the transmission is handed back over to the PSTN for delivery to the other party. The earliest ITSPs have been mostly long-distance resellers seeking the least expensive routes in order to offer their customers the best possible rates. The savings can be substantial, particularly for international calling. See also **IP telephony**.

internetwork connection See **gateway.**

internetworking The process of interconnecting, usually by way of a **gateway**, two or more individual networks to facilitate communications among their respective nodes. The interconnected networks may be different types. Each network is distinct, with its own addresses, internal protocols, access methods, and administration.

interoffice trunk A single direct transmission channel, *e.g.,* voice-frequency circuit, between central offices.

interoperability The ability of systems to provide services to and accept services from other systems and to use the services so exchanged to enable them to operate effectively together.

interoperability standard A document that establishes engineering and technical requirements that are necessary to be employed in the design of systems and to use the services so exchanged to enable them to operate effectively together.

interswitch trunk A single direct transmission channel, *e.g.*, voice-frequency circuit, between switching nodes.

intra-LATA Within the boundaries of a local access and transport area (LATA).

intraoffice trunk A single direct transmission channel, *e.g.*, voice-frequency circuit, within a given switching center.

intranet A private computer network limited to one company or organization. It can be limited to one building or spread out across the country, using high-speed leased lines provided by the phone company. Access is carefully controlled, and the network is often protected by a **firewall**, which makes it difficult for hackers to gain access. As the Internet became more widely used, networking professionals realized that many people were becoming familiar with browsers and Web sites, and these tools are now used in intranets too. Intranets have significantly benefited companies by making it much easier for employees in different departments to share information, for example, or to access information about employment benefits. An intranet makes exchanging information between far-flung offices virtually instantaneous and very inexpensive. Most intranets are connected to the Internet through secure gateways that protect the intranet from intruders from outside and control and monitor Internet usage by employees. When two or more companies or organizations link their intranets, this larger network is referred to as an **extranet**.

Inward Wide-Area Telephone Service (INWATS) See **eight hundred (800) service.**

IP Abbreviation for **information provider**, intelligent peripheral, **Internet protocol**.

IP telephony Technology that allows voice to be transmitted over Internet protocol (IP) data networks, bypassing the public switched telephone network (PSTN) either entirely or for part of the transmission path. The transmission can take three forms, including 1) computer to computer, 2) computer to telephone, or 3) telephone to telephone. The methods utilizing a regular analog telephone at one or both ends requires the services of an **Internet telephony service provider** (ITSP) to serve as the gateway between the PSTN and the data network (usually the Internet).

The caller's voice is packetized into data packets using Internet protocol for transmission over the network. Some organizations are using IP telephony over their WANs, saving money on long-distance and international calls. This is the purest form of IP telephony because the PSTN is totally bypassed. The telephone instrument itself performs the pulse code modulation (PCM), converting analog voice sound waves into digital signals, and plugs directly into the LAN via an RJ-45 jack. Theoretically, the voice quality is superior to a PSTN call because there is never an analog component to the transmission. It's basically free (after buying the equipment) because the voice traffic is riding on an existing data network. Also called *voice over IP (VoIP).*

ISDN See **integrated services digital network.**

ISO See **International Organization for Standardization**.

ITU See **International Telecommunication Union.**

ITU-R The Radiocommunications Sector of the ITU; responsible for studying technical issues related to radiocommunications and having some regulatory powers. A predecessor organization was the CCIR.

ITU-T Abbreviation for **International Telecommunication Union — Telecommunication Standardization Bureau.** The Telecommunications Standardization Sector of the International Telecommunication Union (ITU). ITU-T is responsible for studying technical, operating, and tariff questions and issuing recommendations on them, with the goal of standardizing telecommunications worldwide. The ITU-T combines the standards-setting activities of the predecessor organizations formerly called the International Telegraph and Telephone Consultative Committee (CCITT) and the International Radio Consultative Committee (CCIR).

IVDT Abbreviation for **integrated voice data terminal.** See **integrated station.**

IXC See **interexchange carrier**.

J

jitter The variation in the delay, or **latency**, of data packets going through a network. This is often caused by the data packets taking different paths through the network to reach the destination. This does not create a problem with one-way data communications, such as e-mail, but can create a problem with real-time voice and video over packet-switched networks.

jumper Connections between terminal blocks on the two sides of a distribution frame, or between terminals on a terminal block. Also called *cross-connect*.

junction point See **node**.

K

key **1.** The button on a telephone set. **2.** Information (usually a sequence of random or pseudo-random binary digits) used initially to set up and periodically change the operations performed in crypto-equipment for the purpose of encrypting or decrypting electronic signals.

keyboard An input device used to enter data by manual depression of keys, which causes the generation of the selected code element sent to the attached computer.

key pulsing A system of sending telephone calling signals in which the digits are transmitted by operation of a pushbutton key set. The type of key pulsing commonly used by users and PBX operators is dual-tone multifrequency (DTMF) signaling. Each pushbutton causes generation of a unique pair of tones.

key set A multiline or multifunction user terminal device. See **key telephone**

system.

key service unit (KSU) See **key telephone system**.

key telephone system (KTS) A telephone system for smaller organizations that provides immediate access from all terminals to outside phone lines and other system features, such as intercom and paging, without attendant assistance. Consists of a key service unit (KSU)—which houses the power supply, central processing unit and computer cards that control the system's optional features—and proprietary key set telephones. Each telephone has buttons for accessing all available telephone lines and system features, so that incoming calls can be picked up by any station. Uses a star, or home-run, wiring scheme where each station is wired directly back to the KSU.

kHz See **kilohertz.**

kilohertz (kHz) A unit of frequency denoting one thousand (10^3) Hz.

KSU See **key service unit**.

KSU-less telephone system Instead of having a KSU cabinet that controls the phones, all the intelligence and electronics are built into each telephone set, and all outside lines (usually no more than 4 lines) are wired into each telephone set. The internal wiring can be either series or star, so there is no need to rewire an office for this system if there are sufficient pairs for the number of lines needed. This is an introductory system for small organizations, that can be purchased off the shelf in electronics stores, and does not have the capabilities and flexibility of key systems or PBX systems.

L

LAN See **local area network**.

land line A colloquial name for conventional telephone facilities. Land lines include conventional twisted-pair lines, fiber-optic cable, carrier facilities, and microwave radio facilities for supporting a conventional telephone channel, but do not include satellite links or mobile telephone links using radio transmissions, which are referred to as *cellular* or *wireless* communications.

LATA See **local access and transport area**.

latency The time delay for a transmission to reach its destination traveling over a communications network, expressed in milliseconds (ms). The maximum generally accepted latency for the voice telephone network is 100 ms.

leased circuit Dedicated common-carrier facilities and channel equipment used by a network to furnish exclusive private line service to a specific user or group of users. Also called *leased line.*

least cost routing See **automatic route selection (ARS).**

LEC See **local exchange carrier.**

leg **1.** A segment of an end-to-end route or path, such as a path from user to user via several networks and nodes within networks. Examples of legs are several sequential microwave links between two switching centers and a transoceanic cable between two shore communications facilities, each connected to a node in a national network. **2.** A connection from a specific node to an addressable entity, such as a communication link from a computer workstation to a hub.

legacy (system or equipment) Existing systems or equipment already owned and in place.

LF See **low frequency**.

line conditioning Services performed by the telephone company to reduce noise, attenuation, and distortion on a circuit, usually the local loop. See also **conditioned circuit** and **conditioned loop**.

line-of-sight (LOS) propagation Wireless electromagnetic or RF propagation in which the direct path from the transmitter to the receiver is a straight, unobstructed line. The need for LOS propagation is most critical at VHF and higher frequencies.

line side The portion of a device that is connected to external, *i.e.,* outside plant, facilities such as trunks, local loops, and channels.

link **1.** The communications facilities between adjacent nodes of a network. **2.** A portion of a circuit connected in tandem with, *i.e.*, in series with, other portions. **3.** A radio path between two points, called a radio link. **4.** In communications, a general term used to indicate the existence of communications facilities between two points. **5.** A conceptual circuit, *i.e.*, logical circuit, between two users of a network, that enables the users to communicate, even when different physical paths are used. In all cases, the type of link, such as data link, downlink, duplex link, fiber-optic link, line-of-sight link, point-to-point link, radio link and satellite link, should be identified. A link may be simplex, half-duplex, or duplex. **6.** In a computer program, a part, such as a single instruction or address, that passes control and parameters between separate portions of the program. **7.** In hypertext, the logical connection between discrete units of data.

loading **1.** The insertion of impedance into a circuit to change the characteristics of the circuit. **2.** In multichannel communications systems, the insertion of white noise or equivalent dummy traffic at a specified level to simulate system traffic and thus enable analysis of system performance. **3.** In telephone systems, the load, *i.e.*, power level, imposed by the busy-hour traffic.

loading coil A coil that does not provide coupling to any other circuit, but is inserted in a circuit to increase its inductance. Loading coils inserted periodically in a pair of wires reduce the attenuation at the higher voice frequencies up to the cutoff frequency of the low-pass filter formed by (a) the inductance of the coils and distributed inductance of the wires, and (b) the distributed capacitance between the wires. Above the cutoff frequency, attenuation increases rapidly. A common application of loading coils is to improve the voice-frequency amplitude response characteristics of twisted-cable pairs.

local access and transport area (LATA) Under the terms of the Modification of Final Judgment (MFJ), a geographical area that generally conforms to standard metropolitan and statistical areas (SMSAs), within which a divested Regional Bell Operating Company (RBOC) is permitted to offer exchange telecommunications and exchange access services. Under the terms of the MFJ, the RBOCs are generally prohibited from providing services that originate in one LATA and terminate in another, *i.e.*, long-distance services, which are carried by interexchange carriers (IXCs).

local area network (LAN) A network of linked computers and peripheral devices (such as modems, printers, and CD-ROMs). Usually limited to individual homes, offices, or buildings, or small geographic areas such as a campus or a commercial complex. LAN simply refers to the size of the network, not the technology that makes it work. A LAN can be linked to the public switched telephone network and/or to the Internet. A LAN enables a group of users to share access to databases, computer applications or programs, printers and modems. PCs linked to the LAN, called *clients*, can access powerful computers, called *servers*. The server stores databases and runs applications so users with less-powerful PCs can access stored information in databases or use application programs that might not fit in their computers' memories. Even if the resources would fit on users' PCs, the LAN is still beneficial because it eliminates the need for duplicate databases, applications, or peripheral devices. The primary protocols (methods) for running a LAN are Ethernet, token ring, fiber distributed data interface (FDDI) and ARCNET, with Ethernet and token ring the most common. Sometimes the term *LAN* is used only for very small networks and the term *campus area network* is used for larger networks. A linked group of LANs within a city is called a metropolitan area network (MAN), while the largest networks, linking cities, states or nations, are called wide-area networks (WANs).

local call **1.** Any call using a single switching facility. **2.** Any call for which an additional charge, *i.e.*, toll charge, is not made to the calling or called party. Calls such as those via 800 numbers do not qualify as local calls, because the called party is charged.

local exchange See **central office.**

local exchange carrier (LEC) This is the local telephone company that provides service within each LATA, including Regional Bell Operating Companies (RBOCs) and independent LECs such as General Telephone (GTE). There are also several hundred small independent LECs that serve less populated rural areas. LECs can be distinguished as *incumbent* local exchange carriers (ILECs), the established phone companies, and *competitive* local exchange carriers (CLECs), new local competitors.

local line See **loop.**

local loop See **loop.**

local measured service (LMS) Instead of a flat rate for unlimited local telephone calls, local measured service is a variable charge based on the number of calls, duration, distance, time of day, or any combination of these variables, in accordance with the local phone company's tariff.

local multipoint distribution service (LMDS) A high-capacity wireless one- and two-way communications technology. LMDS can support a wide range of communication services for home and business use, including telephone service, high-speed Internet access, videoconferencing, interactive video, video-on-demand, and remote LAN access. "Because of its multi-purpose applications, LMDS has the potential to become a major competitor to local exchange and cable television services," according to the Federal Communications Commission (FCC). Initial LMDS discussions centered around video transmission, but high-speed data seems to be the greatest application today. In an auction of spectrum in early 1998 similar to those for other wireless services the FCC sold 864 LMDS licenses to bidders interested in building services based on this wireless transmission technology. To encourage competition, the FCC sold two licenses to offer LMDS services in each of 493 metropolitan areas. The auction generated more than a half billion dollars; the FCC retained 122 licenses. LMDS uses frequencies ranging from 27,500 megahertz (27.5 gigahertz) to 31,300 megahertz (31.3 gigahertz). Since consumers and corporations have developed a ravenous appetite for bandwidth, LMDS looks like a promising way to provide a huge amount of bandwidth for little initial investment. Until recently, other forms of broadband communications, such as hybrid fiber/coax, switched broadband and digital broadcast satellites, had received more attention, but LMDS has become more feasible with the maturation of several supporting technologies. Some Third World countries, anxious to improve their telecommunications infrastructure, are considering LMDS as a quicker, more inexpensive alternative to cable or fiber optics. LMDS is currently limited to line-of-sight (LOS) propagation (a clear, straight path between transmitter and receiver), and the signals are degraded by rain and other precipitation.

local number portability (LNP) The ability for subscribers to keep the same telephone number when changing carriers, specifically when changing from the incumbent local exchange carrier (ILEC) to a competitive local exchange carrier (CLEC), as mandated by the Telecommunications Act of 1996.

local office See **central office.**

long-distance call Any telephone call to a destination outside the local service area of the calling station, whether inter-LATA or intra-LATA, and for which there is a charge beyond that for basic service. Also termed **toll call.**

long-haul communications In public switched networks, pertaining to circuits that span large distances, such as the circuits in inter-LATA, interstate, and international communications.

loop **1.** A communications channel from a switching center or an individual message distribution point to the user terminal. Also called **subscriber line.** **2.** In telephone systems, a pair of wires from a central office to a subscriber's telephone. Also called *local loop, local line.*

loop start A supervisory signal given by a telephone or PBX in response to completing the loop path.

LOS See line-of-sight propagation.

low frequency (LF) Any frequency in the band from 30 kHz to 300 kHz.

M

main distribution frame (MDF) A distribution frame on one side of which the external trunk cables entering a facility terminate, and on the other side of which the internal user subscriber lines and trunk cabling to any intermediate distribution frames terminate. The MDF is used to cross-connect any outside line with any desired internal cabling or any other outside line. The MDF usually holds central office protective devices and functions as a test point between a line and the office. The MDF in a private exchange (PBX) performs functions similar to those performed by the MDF in a central office. Also called **main frame.**

make set busy A service feature that busies out your phone without having to take it off-hook. Especially useful for the those who need to ignore incoming calls to complete a project.

MAN See **metropolitan area network.**

MAPI See **messaging application programming interface**.

master station **1.** In a data network, the station that is designated by the control station to ensure data transfer to one or more slave stations. A master station controls one or more data links of the data communications network at any given instant. The assignment of master status to a given station is temporary and is controlled by the control station according to the procedures set forth in the operational protocol. Master status is normally conferred upon a station so that it may transmit a message, but a station need not have a message to send to be designated the master station. **2.** In basic mode link control, the data station that has accepted an invitation to ensure a data transfer to one or more slave stations. At a given instant, there can be only one master station on a data link.

MDF See **main distribution frame.**

measured-rate service Telephone service for which charges are made in accordance with the total connection time of the line, distance between parties, number of calls, time of day, or any combination of such variables. See also **local measured service**.

medium frequency (MF) Frequencies from 300 kHz to 3000 kHz.

megahertz (MHz) A unit of frequency denoting one million (10^6) Hz.

message handling system (MHS) In the CCITT X.400 Recommendations, the family of services and protocols that provides the functions for global electronic-mail transfer among local mail systems.

message service Switched service furnished to the general public (as distinguished from private line service). Except as otherwise provided, this includes exchange switched services and all switched services provided by interexchange carriers and completed by a local telephone company's access services. Also called **message toll service.**

message toll service See **message service**.

message unit A unit of measure for charging telephone calls, based on parameters such as the length of the call, the distance called, and/or the time of day.

messaging application programming interface (MAPI) Microsoft's Windows-based application programming interface (API) that facilitates the transfer of messages (*i.e.*, e-mail, fax) among users via various messaging platforms and service providers.

messaging service In integrated services digital networks (ISDN), an interactive telecommunications service that provides for information interchange among users by means of store-and-forward, electronic mail, or message-handling functions.

metropolitan area network (MAN) A network, larger than a local area network (LAN) but smaller than a wide area network (WAN), created through the connection of smaller networks. A MAN often refers to the interconnection of networks within a city to create a larger network, but not always. The term can also be used to describe a network created by connecting two or more LANs. The wireless, cell-switched service Switched Multimegabit Data Service (SMDS), was designed for interconnecting LANs in a MAN.

MF See **medium frequency**.

MFJ See **Modification of Final Judgment**.

MHz See **megahertz**.

microwave (mw) Loosely, an electromagnetic wave having a wavelength from 300 mm to 10 mm (1 GHz to 30 GHz). Microwaves exhibit many of the properties usually associated with waves in the optical regime, *e.g.*, they are easily concentrated into a beam.

microwave multipoint distribution system (MMDS) A wireless, line-of-sight (LOS) cable TV technology that replaces coaxial cable wiring to every subscriber. It consists of a transmitter that broadcasts to multiple receivers within a 40-mile radius, as long as there are no obstructions in the LOS transmission path. See also **local multipoint distribution service (LMDS)**.

MIS Abbreviation for management information system.

mobile services switching center (MSC) In an automatic cellular mobile system, the interface between the radio system and the public switched telephone network. The MSC performs all signaling functions that are necessary to establish calls to and from mobile stations.

modem Acronym for **modulator/demodulator.** A device usually attached to a computer that converts data back and forth between digital and analog formats to be sent over the analog local portion of the public telephone network. Since computers use a digital format to store and process data, the binary numbers (series of zeroes and ones) that comprise the data must be changed into the fluctuating electrical signals of an analog transmission. At the receiving end, another computer's modem translates the analog signal back into digital data. Most modems include error detection and correction capabilities as well as internal dialing. The fastest PC modems operate at 56 Kbps and have not yet been standardized. The terminal adapters (TAs) used in ISDN environments also can be called modems, as can the terminating units (TUs) used with ADSL.

Modification of Final Judgment (MFJ) The 1982 antitrust suit settlement agreement ("Consent Decree") entered into by the United States Department of Justice and the American Telephone and Telegraph Company (AT&T) that, after modification and upon approval of the United States District Court for the District of Columbia, required the divestiture of the Bell Operating Companies from AT&T.

modular jack A device that conforms to the *Code of Federal Regulations,* Title 47, part 68, which defines the size and configuration of all units that are permitted for connection to the public exchange facilities.

modulation The process, or result of the process, of varying a characteristic of a carrier, in accordance with an information-bearing signal that modulates, or modifies, the carrier signal. In analog telecommunications, the carrier is an electrical signal, analogous to a sound signal, that is represented by a sine wave with constant amplitude and frequency. The characteristics of the carrier signal that can be modified are its amplitude, frequency, or phase.

modulation rate The rate at which a carrier is varied to represent the information in a digital signal. Modulation rate and information transfer rate are not necessarily the same.

Mosaic A portable World Wide Web browser that provides a graphical user interface to hypertext-based information.

MPLS See **multiprotocol label switching.**

MSO See **multiple system operator**.

multicarrier modulation (MCM) A technique of transmitting data by dividing the data into several interleaved bit streams and using these to modulate several carriers. MCM is a form of **frequency-division multiplexing**.

multicast **1.** In a network, a technique that allows data, including packet form, to be simultaneously transmitted to a selected set of destinations. Some networks, such as Ethernet, support multicast by allowing a network interface to belong to one or more multicast groups. **2.** To transmit identical data simultaneously to a selected set of destinations in a network, usually without obtaining acknowledgement of receipt of the transmission.

multicast address A routing address that (a) is used to address simultaneously all the computers in a group and (b) usually identifies a group of computers that share a common protocol, as opposed to a group of computers that share a common network. Multicast address also applies to radio communications. Also called (in Internet protocol) *class d address*.

multichannel Pertaining to communications, usually full-duplex, on more than one channel. Multichannel transmission may be accomplished by time-division multiplexing, frequency-division multiplexing, phase-division multiplexing, or space diversity.

multichannel multipoint distribution system (MMDS) A method of delivering cable TV to subscribers using microwave signals in the 2.5 GHz to 2.7 GHz band. MMDSs consist of transmission antennas that deliver more than 100 channels to subscribers. Because microwaves cannot penetrate solid objects, like hills or buildings, subscribers' receiving dishes must be within 40 miles and on a clear "line of sight" from the transmitter. MMDS is being used in developing countries because it is relatively inexpensive to serve a large number of subscribers. MMDS, in part because it is a broadband digital system and also because it uses MPEG-2 compression, can carry more than three times as many cable stations as analog systems. MPEG (Moving Pictures Experts Group—an International Standards Organization group) is a set of standards for compressing full-motion video. MPEG-2 is becoming a key compression scheme for digital video and broadcast transmissions.

multifiber cable A fiber-optic cable having two or more fibers, each of which is capable of serving as an independent optical transmission channel.

multilink operation In packet-switched networks, the simultaneous use of multiple links for the transmission of different segments of the same message unit. Use of multilink operation is intended to increase the effective rate of message transmission. Multilink operation requires special procedures for multiplexing/demultiplexing control.

multimedia Pertaining to the processing and integrated presentation of information in more than one form, *e.g.*, video, voice, music, or data.

multipath The propagation phenomenon that results in radio signals' reaching the receiving antenna by two or more paths. Causes of multipath include atmospheric ducting, ionospheric reflection and refraction, and reflection from terrestrial objects, such as mountains and buildings. The effects of multipath include constructive and destructive interference, and phase shifting of the signal. In facsimile and television transmission, multipath causes jitter and ghosting.

multiple access 1. The connection of a user to two or more switching centers by separate access lines using a single message routing indicator or telephone number. 2. In satellite communications, the capability of a communications satellite to function as a portion of a communications link between more than one pair of satellite terminals concurrently. The three types of multiple access presently used with communications satellites are code-division, frequency-division, and time-division multiple access. 3. In computer networking, a scheme that allows temporary access to the network by individual users, on a demand basis, for the purpose of transmitting information. Examples of multiple access are carrier sense multiple access with collision avoidance (CSMA/CA) and carrier sense multiple access with collision detection (CSMA/CD).

multiple frequency-shift keying (MFSK) A modulation technique in which multiple codes are used in the transmission of digital signals. In MFSK, the coding schemes use multiple frequencies that are transmitted concurrently or sequentially. See also **frequency-shift keying**.

multiple homing In telephone systems, the connection of a terminal facility so that it can be served by one or several switching centers, or the connection of a terminal facility to more than one switching center by separate access lines. Separate directory numbers are applicable to each switching center accessed.

multiple media Transmission media using more than one type of transmission path (*e.g.*, optical fiber, radio, and copper wire) to deliver information. Contrast with **multimedia.**

multiple system operator (MSO) A company that operates more than one cable TV system.

multiplex A method of sending two or more messages on a single communications channel by using a device called a **multiplexer** to combine the individual signals at the transmitting end of the channel and divide them at the receiving end. For analog signals, the channel is shared using frequency-division multiplexing (FDM), while digital signals share a channel using time-division multiplexing (TDM). Satellite, land-line and cable system signals all can be multiplexed. In optical communications, the analog of FDM is referred to as wavelength-division multiplexing (WDM).

multiplexer (MUX) A device that combines multiple inputs into an aggregate signal to be transported via a single transmission channel. See **multiplex**.

multiplexer/demultiplexer (muldem) A device that combines the functions of multiplexing and demultiplexing of digital signals. The term *muldem* should not be confused with *modem*.

multiplex hierarchy In frequency-division multiplexing, the rank of frequency bands occupied:

12 channels	group
5 groups (60 channels)	super group
5 super groups (300 channels)	master group (CCITT)
10 super groups (600 channels)	master group (U.S. standard)
6 U.S. master groups (3600 channels)	jumbo group

multiport repeater In digital networking, an active device having multiple input/output (I/O) ports, in which a signal introduced at the input of any port appears at the output of every port. A multiport repeater usually performs regenerative functions, *i.e.*, it reshapes the digital signals. Depending on the application, a multiport repeater may be designed not to repeat a signal back to the port from which it originated.

multiprotocol label switching (MPLS) This is a new networking protocol that offers better quality of service (QoS) for high priority data traffic. The conventional Internet protocol (IP) simply addresses each data packet individually and sends it on its way through the network without regard to its relative importance, and the data packets for any given session can take multiple paths traveling from the sender to the receiver. Instead, with MPLS, all packets in an IP session are grouped together into a single "flow" and then tagged for priority handling through the network routers. The group of packets, or flow, is mapped onto a dedicated path in an ATM or frame relay virtual circuit, a SONET channel, or any other data link (OSI layer 2) for delivery. MPLS mapping is actually accomplished by swapping layer 3 labels for layer 2 labels in a data packet header, hence the term *label switching*. In this manner, data transmissions can be assigned different levels of priority so that financial file transfers, for example, get higher priority than e-mail.

MUX See **multiplexer**.

mw See **microwave.**

N

NAP See **network access point.**

narrowband signal Any analog signal or analog representation of a digital signal whose essential spectral content is limited to that which can be contained within a voice channel of nominal 4-kHz bandwidth. Narrowband radio uses a voice channel with a nominal 3-kHz bandwidth.

National Information Infrastructure (NII) A proposed, advanced, seamless web of public and private communications networks, interactive services, interoperable hardware and software, computers, databases, and consumer electronics to put vast amounts of information at users' fingertips. Also referred to as the *information superhighway*. NII includes more than just the physical facilities (cameras, scanners, keyboards, telephones, fax machines, computers, switches, compact disks, video and audio tape, cable, wire, satellites, optical fiber transmission lines, microwave nets, switches, televisions, monitors, and printers) used to transmit, store, process, and display voice, data, and images; it encompasses a wide range of interactive functions, user-tailored services, and multimedia databases that are interconnected in a technology-neutral manner that will favor no one industry over any other.

negative acknowledge character (NAK) A data transmission control character sent by a station as a negative response to the station with which the connection has been set up. In binary synchronous communication protocol, the NAK is used to indicate that an error was detected in the previously received block and that the receiver is ready to accept retransmission of that block. In multipoint systems, the NAK is used as the *not-ready* reply to a poll.

network An interconnection of three or more communicating entities. This glossary covers telephone, cellular, and data (computer) networks, for example.

network access point (NAP) A point on the Internet where networks and service providers hand off traffic to each other. For example, two or more major Internet backbone carriers would interconnect at a NAP, along with a host of smaller Internet service providers (ISPs), all swapping traffic back and forth using a variety of **peering** arrangements. NAPs are busy interconnection points where the worst congestion occurs on the Internet.

network administration A group of network management functions that (a) provide support services, (b) ensure that the network is used efficiently, and (c) ensure prescribed service quality objectives are met. Network administration may include activities such as network address assignment, assignment of routing protocols and routing table configuration, and directory service configuration.

network architecture The design principles, physical configuration, functional organization, operational procedures, and data formats used as the basis for the design, construction, modification, and operation of a communications network. Networks can be classified in terms of the switching mechanism they use (circuit-switched, packet-switched), the data transmission protocols they use (TCP/IP, frame relay, ATM, etc.) or their size (LANs, MANs, WANs).

A circuit-switched network designates, for the duration of a communication, a single physical path between two points on the network. Voice telephone service is circuit-switched and uses time-division multiplexing to control access to the network. In contrast, a packet-switched network allows users to share a data communications channel, or circuit, thus optimizing the use of network resources. Ethernet and token ring LANs both are examples of packet-switched networks that use statistical multiplexing to share capacity and control access.

Cell-switching is the newest form of packet-switching. A packet can vary in length, based on a number of factors, while a cell is always the same length. Asynchronous transfer mode (ATM) is an example of a cell-switching technology. Also considered a network architecture, Open Systems Interconnection (OSI), is actually a reference model for a network structure. Individual manufacturers and service providers can build equipment and networks based on this model, which helps ensure interoperability between individual components of a network as well as between multiple networks.

According to the Insight Research Corporation, a firm specializing in telecommunications market research, analysis and consulting, no single network architecture today can provide a "complete" communications solution. Service providers will continue to offer alternative services, based on different network architectures, to meet the needs of different markets, such as telephony, cable TV, wireless, and so on. Insight predicts that carriers will mix-and-match network architectures to align the costs and benefits of a particular architecture with the requirements of each group of customers.

network computer (NC) A stripped-down personal computer that relies on a network (typically a LAN) to deliver information from a central server (a powerful computer) that stores applications, databases, software, etc. An NC is inexpensive because it does not include a disk drive, CD-ROM drive, or expansion slots. Some are devoid of modems, ports, and hard drives. The NC is appealing to some businesses because it greatly reduces the cost of a PC. Oracle and Sun Microsystems have been the leading proponents of the network computer. The network computer can also be called a thin *client*, and is similar to a NetPC. A NetPC is a thin client that is connected to the Internet and relies on servers maintained by an Internet Access Provider for access to applications like word processing, for example, as well as storage space for personal files and even processing power to drive the NetPC.

network control program (NCP) In a switch or network node, software designed to store and forward frames between nodes. An NCP may be used in local area networks or larger networks.

network element A facility or equipment used in the provision of a telecommunications service. Such term also includes, but is not limited to, features, functions, and capabilities that are provided by means of such facility or equipment, including but not limited to, subscriber numbers, databases, signaling systems, and information sufficient for billing and collection or used in the transmission, routing, or other provision of a telecommunications service.

network interface **1.** The point of interconnection between a user terminal and a private or public network. **2.** The point of interconnection between the public switched network and a privately owned terminal. The *Code of Federal Regulations,* Title 47, part 68, stipulates the interface parameters. **3.** The point of interconnection between one network and another network.

network interface card (NIC) A network interface device (NID) in the form of a circuit card that is installed in an expansion slot of a computer to provide network access. Examples of NICs are cards that interface a computer with an Ethernet LAN and cards that interface a computer with an FDDI ring network.

network interface device (NID) **1.** A device that performs interface functions, such as code conversion, protocol conversion, and buffering, required for communications to and from a network. **2.** A device used primarily within a local area network (LAN) to allow a number of independent devices, with varying protocols, to communicate with each other. An NID converts each device protocol into a common transmission protocol. The transmission protocol may be chosen to accommodate directly a number of the devices used within the network without the need for protocol conversion for those devices by the NID. Also called **network interface unit.**

network interface unit (NIU) See **network interface device**.

network management The execution of the set of functions required for controlling, planning, allocating, deploying, coordinating, and monitoring the resources of a telecommunications network, including performing functions such as initial network planning, frequency allocation, predetermined traffic routing to support load balancing, cryptographic key distribution authorization, configuration management, fault management, security management, performance management, and accounting management. Network management does not include user terminal equipment.

network operating system (NOS) Software that (a) controls a network, its message (*e.g.*, packet) traffic, and queues, (b) controls access by multiple users to network resources such as files, and (c) provides for certain administrative functions, including security. A network operating system is most frequently used with local area networks and wide area networks, but could also have application to larger network systems.

network terminal number (NTN) In the CCITT International X.121 format, the sets of digits that comprise the complete address of the data terminal end point. For an NTN that is not part of a national integrated numbering format, the NTN is the 10 digits of the CCITT X.25 14-digit address that follow the Data Network Identification Code (DNIC). When part of a national integrated numbering format, the NTN is the 11 digits of the CCITT X.25 14-digit address that follow the DNIC.

network terminating interface (NTI) See **demarcation point.**

network termination 1 (NT1) In Integrated Services Digital Networks (ISDN), a functional grouping of customer premises equipment (CPE) that includes functions that may be regarded as belonging to OSI Layer 1, *i.e.,* functions associated with ISDN electrical and physical terminations on the user premises. The NT1 forms a boundary to the network and may be controlled by the provider of the ISDN services. The NT1 can be built into ISDN CPE (i.e., ISDN telephones) or be a stand-alone device into which ISDN CPE devices are connected.

network termination 2 (NT2) In Integrated Services Digital Networks (ISDN), an intelligent device that may include functionality for OSI Layers 1 through 3 (dependent on individual systems requirements).

network topology The specific physical, *i.e.,* real, or logical, arrangement of the elements of a network. Two networks have the same topology if the connection configuration is the same, although the networks may differ in physical interconnections, distances between nodes, transmission rates, and/or signal types. The common types of network topology are illustrated and defined in alphabetical order below:

bus topology: A network topology in which all nodes, *i.e.,* stations, are connected together by a single bus.

fully connected topology: A network topology in which there is a direct path (branch) between any two nodes. In a fully connected network with *n* nodes, there are n(n–1)/2 direct paths, *i.e.,* branches. Also called *fully connected mesh network* or *mesh network.*

hybrid topology: A combination of any two or more network topologies. Instances can occur where two basic network topologies, when connected together, can still retain the basic network character, and therefore not be a hybrid network. For example, a tree network connected to a tree network is still a tree network. Therefore, a hybrid network forms only when two basic networks are connected and the resulting network topology fails to meet one of the basic topology definitions. For example, two star networks connected together exhibit hybrid network topologies. A hybrid topology always happens

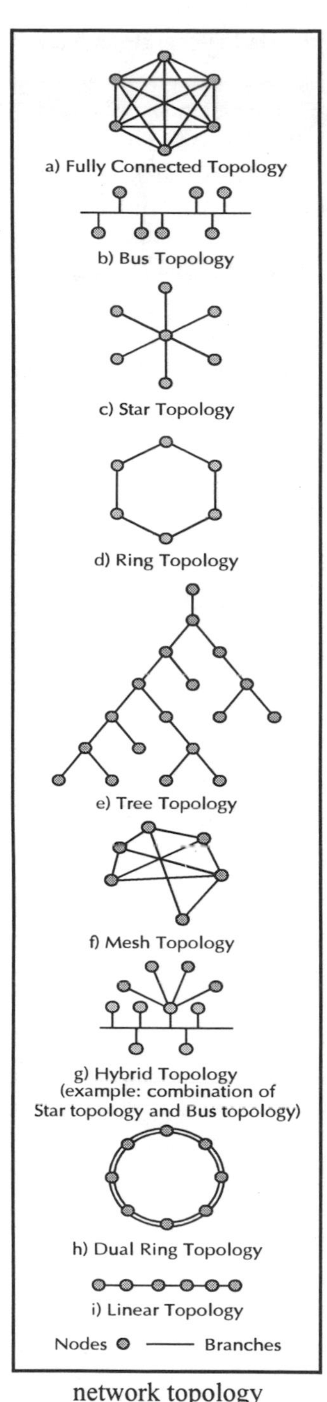

network topology

when two different basic network topologies are connected.

linear topology: *See* **bus topology.**

mesh topology: A network topology in which there are at least two nodes with two or more paths between them.

ring topology: A network topology in which every node has exactly two branches connected to it.

star topology: A network topology in which peripheral nodes are connected to a central node, which rebroadcasts all transmissions received from any peripheral node to all peripheral nodes on the network, including the originating node. All peripheral nodes may thus communicate with all others by transmitting to, and receiving from, the central node only.

tree topology: A network topology that, from a purely topological viewpoint, resembles an interconnection of star networks in that individual peripheral nodes are required to transmit to and receive from one other node only, toward a central node, and are not required to act as repeaters or regenerators. The function of the central node may be distributed.

Next Generation Network (NGN) The network of the future. Although this is still more concept than reality, most experts agree that NGN will be a high-speed packet-switched network that merges data, voice, video, and multimedia. NGN is seen as the merging of the PSTN and the Internet into one common global network that does everything.

NIC See **network interface card.**

NID See **network interface device.**

NII See **National Information Infrastructure.**

nine hundred (900) service A telephone service via which the caller may access information or entertainment services on a charge-per-call or charge-per-time basis, at a rate in excess of the underlying transport (toll) charges. The entity providing the service, the information provider (IP), can set the charges at a level to cover all transport and other costs plus profit.

node In a network, the term *node* typically refers to a device that can forward transmissions to another node on the network. It is usually located at a point where two branches of the network meet. A node on a switched network can be one of the switches that forms the network's backbone. Each node on a network has a unique address.

North American Numbering Plan (NANP) The method of identifying telephone trunks and assigning service access codes (area codes) in the public network of North America, including the U.S., Canada and the Caribbean. New service access codes are assigned by the North American Numbering Plan Administration (NANPA). See also **area code**.

NXX code In the North American direct-distance dialing numbering plan, a central office code of three digits that designates a particular central office or a given 10,000-line unit of subscriber lines; "N" is any number from 2 to 9, and "X" is any number from 0 to 9.

O

octet A byte of eight binary digits usually operated upon as an entity.

off-hook In telephony, the condition that exists when an operational telephone instrument or other user instrument is in use, either dialing or communicating. Off-hook originally referred to the condition that prevailed when the separate earpiece, (receiver) was removed from its **switchhook**, which extended from a vertical post that also supported the microphone, and which also connected the instrument to the line when not depressed by the weight of the receiver.

off-hook signal In telephony, a signal indicating seizure, request for service, or a busy condition.

office classification Prior to divestiture, numbers were assigned to switching offices according to their hierarchical function in the U.S. public switched telephone network. The following class numbers are used:

Class 1: Regional Center (RC)
Class 2: Sectional Center (SC)
Class 3: Primary Center (PC)
Class 4: Toll Center (TC) [Only if operators are present; otherwise Toll Point (TP)]
Class 5: End Office (EO) [Local central office]
Note: Any one center handles traffic from one center to two or more centers lower in the hierarchy. Since divestiture, these designations have become less firm.

off-premises extension (OPX) An extension telephone, PBX station, or key system station located on property that is not contiguous with that on which the main telephone, PBX, or key system is located.

omnidirectional antenna An antenna that has a radiation pattern that is nondirectional in azimuth. The vertical radiation pattern may be of any shape.

ONA See **open network architecture.**

one-way communication Communication in which information is always transferred in only one preassigned direction. One-way communication is not necessarily constrained to one transmission path. Examples of one-way communications systems include broadcast stations, one-way intercom systems, and wireline news services.

one-way trunk A trunk between two switching centers, over which traffic may be originated from one preassigned location only. The traffic may consist of two-way communications; the expression *one-way* refers only to the origin of the demand for a connection. At the originating end, the one-way trunk is known as an *outgoing trunk;* at the other end, it is known as an *incoming trunk.*

on-hook In telephony, the condition that exists when an operational telephone, or other user instrument, is not in use. On-hook originally referred to the storage of an idle telephone receiver on a hook that extended from a vertical post that supported the microphone also. The hook was mechanically connected to a switch that automatically disconnected the idle telephone from the network.

on-hook signal In telephony, a signal indicating a disconnect, unanswered call, or an idle condition.

on-line call detail data (OCDD) Information summarizing inbound calling data, typically detailing call volumes originating from different telephone area codes or states. Uses automatic number identification (ANI) to compile the information. Useful for tracking total call volume from different geographic areas, perhaps in response to a promotion of some sort.

on-premises extension An extension telephone, PBX station, or key system station located on property that is contiguous with that on which the main telephone, PBX, or key system is located.

on-premises wiring Customer-owned metallic or optical-fiber communications transmission lines, installed within or between buildings. On-premises wiring may consist of horizontal wiring, vertical wiring, and backbone wiring, and may extend from the external network interface to the user work station areas. It includes the total communications wiring to transport current or future data, voice, LAN, and image information.

open circuit In communications, a circuit available for use.

open network architecture (ONA) The overall design of a communication carrier's basic network facilities and services to permit all users, including competitive carriers, to interconnect to specific basic network functions and interfaces on an unbundled, equal-access basis. The ONA concept consists of three integral components: (a) basic serving arrangements (BSAs), (b) basic service elements (BSEs), and (c) complementary network services.

open systems architecture The layered hierarchical structure, configuration, or model of a communications or distributed data processing system that (a) enables system description, design, development, installation, operation, improvement, and maintenance to be performed at a given layer or layers in the hierarchical structure, (b) allows each layer to provide a set of accessible functions that can be controlled and used by the functions in the layer above it, (c) enables each layer to be implemented without affecting the implementation of other layers, and (d) allows the alteration of system performance by the modification of one or more layers without altering the existing equipment, procedures, and protocols at the remaining layers. Examples of independent alterations include (a) converting from copper wire to optical fibers at a physical layer without affecting the data link layer or the network layer except to provide more traffic capacity, and (b) altering the operational protocols at the network level without altering the physical layer. Open systems architecture may be implemented using the Open Systems Interconnection-Reference Model (OSI-RM) as a guide while designing the system to meet performance requirements.

Open Systems Interconnect (OSI) A model of a network architecture built so that different networks and computer systems made by different vendors can communicate with each other. The International Standards Organization (ISO) developed OSI, including the protocols to implement it, so that data can be shared among all systems on the same network or a linked network. Although the suite of protocols, also referred to as the *protocol stack*, for OSI remains a model, many of the leading communications protocols incorporate this model. TCP/IP networks, for example, are based on a five-layer protocol stack. The OSI architecture consists of seven layers. They are: physical layer, data link layer, network layer, transport layer, session layer, presentation layer, and application layer (see below). By dividing the protocols for the network architecture into clearly defined layers, it is possible to specifically define how information will be passed from one layer to the next. It also allows changes to be made to the protocols for one layer without affecting the entire network architecture.

The seven layers of the OSI model are hierarchical. In other words, the higher layer in essence tells the adjacent lower layer what to do and that layer, in turn, tells the next lower layer what service to perform. For example, Layer 1, the physical layer, is the lowest. It establishes and terminates connections, conveying the bit stream through the network, based on requests from Layer 2, the data link layer. The data link layer provides synchronization for the physical layer and can detect and sometimes correct errors in the physical layer. Data link protocols include HDLC (High-level data link control) for packet-switched networks and LLC (Logical link control) for LANs. The data link layer responds to requests from the network layer.

The network layer, which follows orders from the transport layer, decides how data will be routed within a network and between linked networks. The transport layer, which takes orders from the session layer, transfers data between end users and determines whether all packets in a transmission have arrived. The session layer, guided by the presentation layer, manages the dialogue between each end users' applications by controlling the basic communications channel provided by the transport layer. The presentation layer, based on guidance from the application layer, can provide translation services for end users' with different data formats.

The Presentation layer is usually a part of an operating system. And finally, the highest layer, the Application layer, is concerned with the user's view of the network, for example, formatting e-mail messages or performing file transfers.

In general, the higher the layer, the more intelligence and more software controlled. The lower layers are relatively dumb and hardware-oriented, but quite fast. The following summarizes the layers:

Physical layer: Layer 1, the lowest of seven hierarchical layers. The physical layer performs services requested by the data link layer. The major functions and services performed by the physical layer are: (a) establishment and termination of a connection to a communications medium; (b) participation in the process whereby the communication resources are effectively shared among multiple users, *e.g.,* contention resolution and flow control; and, (c) conversion between the representation of digital data in user equipment and the corresponding signals transmitted over a communications channel. Hardware such as hubs and repeaters are found at this layer.

Data link layer: Layer 2. This layer responds to service requests from the network layer and issues service requests to the physical layer. The data link layer provides the functional and procedural means to transfer data between network entities and to detect and possibly correct errors that may occur in the physical layer. Examples of data link protocols are HDLC and ADCCP for point-to-point or packet-switched networks and LLC for local area networks. Hardware such as bridges and switches are found at this layer.

Network layer: Layer 3. This layer responds to service requests from the transport layer and issues service requests to the data link layer. The network layer provides the functional and procedural means of transferring variable length data sequences from a source to a destination via one or more networks while maintaining the quality of service requested by the transport layer. The network layer performs network routing, flow control, segmentation, desegmentation, and error control functions. Routers and brouters are found at this layer.

Transport layer: Layer 4. This layer responds to service requests from the session layer and issues service requests to the network layer. The purpose of the transport layer is to provide transparent transfer of data between end users, thus relieving the upper layers from any concern with providing reliable and cost-effective data transfer. Message integrity is maintained at this layer.

Session layer: Layer 5. This layer responds to service requests from the presentation layer and issues service requests to the transport layer. The session layer provides the mechanism for managing the dialogue between end-user application processes. It provides for either duplex or half-duplex operation and establishes checkpointing, adjournment, termination, and restart procedures. Gateways that join networks of differing protocols are at this layer.

Presentation layer: Layer 6. This layer responds to service requests from the application layer and issues service requests to the session layer. The presentation layer relieves the application layer of concern regarding syntactical differences in data representation within the end-user systems. An example of a presentation service would be the conversion of an EBCDIC-coded text file to an ASCII-coded file.

Application layer: Layer 7, the highest layer. This layer interfaces directly to and performs common application services for the application processes; it also issues requests to the presentation layer. The common application services provide semantic conversion between associated application processes. For example, users at each end would have to use the same word processing software in order to properly exchange files.

operations support systems (OSS) Also known as *operating* support systems; the procedures, equipment, and computer programs that monitor and administer telecommunications networks and infrastructure. These automated management systems support network operations such as billing customers, processing orders, provisioning new services, or testing lines.

operator service provider (OSP) A specialized telecommunications company that provides operator-assisted services for collect calls, credit card calls, and person-to-person calls, usually dialed from private payphones, hotels, or public facilities.

optical carrier level (OC-1 through OC-192) The SONET digital hierarchy that defines data rates from OC-1 at 51.84 Mbps to OC-192 at 9,952 Mbps. See also **SONET**.

optical fiber A filament of transparent dielectric material, usually glass or plastic, and usually circular in cross section, that guides light. An optical fiber usually has a cylindrical core surrounded by, and in direct contact with, a protective cladding of similar geometry.

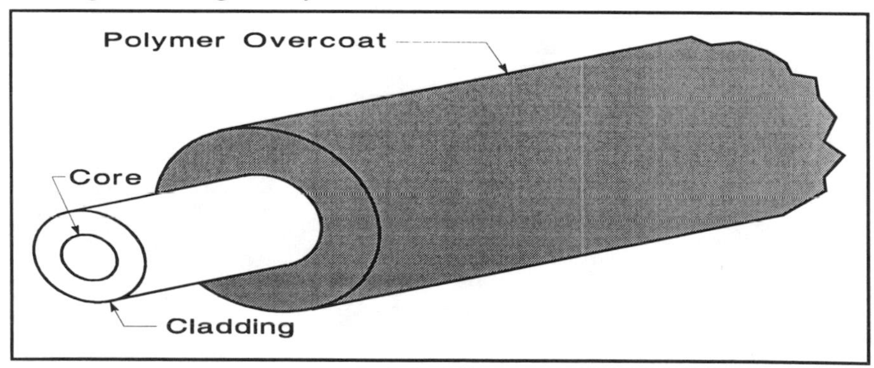

optical fiber

optical fiber cable See **fiber optic cable.**

optical repeater In an optical communication system, an opto-electronic device or module that receives an optical signal, amplifies it (or, in the case of a digital signal, reshapes, retimes, or otherwise reconstructs it), and retransmits it as an optical signal.

out-of-band signaling Special signaling that is carried in a separate channel from the channel that carries the data or information. Supervisory and dialing signals are examples of out-of-band signals that accompany a phone call or data transmission.

outside plant In telephony, all cables, conduits, ducts, poles, towers, repeaters, repeater huts, and other equipment located between a demarcation point in a switching facility and a demarcation point in another switching facility or at the customer premises. Basically, everything outside of the central offices and switching facilities.

overhead In data communications, extra information that is added to the main message transmitted. It assists in addressing, routing, error-checking and otherwise helping to get the data to the intended recipient in a decipherable form.

P

PABX Abbreviation for *private automatic branch exchange*. See **private branch exchange (PBX).** The term *PBX* is more common than *PABX,* regardless of automation, which is now assumed.

packet A unit of data, existing as a group of zeroes and ones, that is a portion of a file or other block of information being transported on a packet-switched network. In a TCP/IP network, the Transport Control Protocol layer divides an entire transmission into smaller pieces—called *packets*—that can be efficiently transported through the network. Each packet contains the destination address, the "payload" (or text being transmitted), and error detection information. Each packet might follow a different route through the network. When all of the packets have arrived at the destination, the TCP layer at the destination reassembles the packets into their original order.

packet assembler/disassembler (PAD) A device that enables data terminal equipment (DTE) not equipped for packet switching to access a packet-switched network. It operates somewhat like a modem between a computer and an analog telephone network, except that it takes the digital computer data and assembles it into packets that can be transmitted over a digital network. Another PAD at the other end disassembles the packets into computer-readable data.

packet format The structure of data, address, and control information in a packet. The size and content of the various fields in a packet are defined by a set of rules that are used to assemble the packet.

packet-switching A method of transporting data that has been broken into small pieces, called *packets*, over shared communications channels. Packet switching optimizes the use of network resources (bandwidth) because the channel is only occupied during the time the packet is being transmitted. In contrast, the regular telephone network keeps an entire channel occupied for the length of a communication, even when there is no immediate activity on the line. On a packet-switched network, many users can share the same channel because individual packets can be sent and received in any order. Examples of packet-switched network protocols include ATM, frame relay, TCP/IP, and X.25. The Internet was the first public packet-switched network.

PAD See **packet assembler/disassembler.**

pager A mobile receiver for paging communications, also known as a *beeper.*

paging A one-way wireless communications service from a base station to mobile or fixed receivers that provide signaling or information transfer by such means as tone, tone-voice, tactile, or optical readout.

part 68 The section of Title 47 of the *Code of Federal Regulations* governing the direct connection of telecommunications equipment and customer premises wiring with the public switched telephone network and certain private line services. Part 68 rules provide the technical and procedural standards under which direct electrical connection of customer-provided telephone equipment, systems, and protective apparatus may be made to the nationwide network without causing harm.

party line In telephone systems, an arrangement in which two or more user end instruments, usually telephones, are connected to the same loop. If selective ringing is not used, individual users may be alerted by different ringing signals, such as a different number of rings or a different combination of long and short rings. Party lines remain primarily in rural areas where loops are long.

passive optical network (PON) A fiber-optic network that uses a minimum of active, electrically-powered deives such as amplifiers; instead relying on passive splitters and combiners that do not require dedicated power sources. The theory is that with fewer "moving parts" in the network it becomes more reliable, less expensive to build, and easier to manage.

patch To connect circuits together temporarily. In communications, patches may be made by means of a cord, *i.e.,* a cable, known as a *patch cord.* In automated systems, patches may be made electronically.

pay-per-call The caller pays a predetermined charge for accessing information or entertainment services, at a rate in excess of the underlying transport charges. 900 numbers are one example of such a service. Pay-per-call services may also be offered over 800 lines or regular toll lines using credit cards or other third-party billing mechanisms. When the caller pays a premium above the regular transport charges for the information content of the program, regardless of how payment is made, it is considered a pay-per-call service.

pay phone Traditionally, coin-operated telephones found in public places. Now they can accept credit cards, so coins are not always needed.

PBX See **private branch exchange**.

PCS See **Personal Communications Service.**

peering The free exchange of data traffic by Internet service providers (ISPs) and carriers at the various **network access points** (NAPs) on the Internet. The NAPs essentially interconnect various Internet carriers and allow for the free flow of traffic among the carriers' networks. Generally, carriers of equal size (similar traffic volume) are willing to carry each other's traffic on their networks because it is an equal trade. The free ride that smaller carriers have been getting by equally sharing networks with the larger carriers may be coming to an end. There is movement towards more equitable private peering arrangements or some kind of settlement system for peering between carriers of different sizes.

peer-to-peer network Local area networks (LANs) are typically peer-to-peer where every node on the network has equal status, able to transmit or receive data at any time without getting permission from a network control node.

permanent virtual circuit (PVC) A virtual circuit used to establish a long-term connection between data terminal equipments (DTE). In a PVC, the long-term association is identical to the data transfer phase of a virtual call. Permanent virtual circuits eliminate the need for repeated call set-up and clearing. Also called *nailed-up circuit.*

Personal Communications Services (PCS) A developing wireless technology that competes with cellular telephony. PCS is a digital service that operates in a higher frequency than regular cellular. PCS telephones are typically smaller, cheaper, and have less range. PCS is more of a walk-around communications service, effective in a city, for example, whereas cellular has a broader range that is effective in motor vehicles. PCS operates in frequencies ranging from 1.5 MHz to 1.8 MHz. In contrast, most cellular systems use the 800 MHz to 900 MHz range. PCS networks use smaller cells, which allows them to use lighter, longer-lasting batteries.

Personal Digital Assistant (PDA) Multi-function communications device that looks like a little palmtop computer, which may perform calendar, memo pad, calculator, and scheduling functions, as well as e-mail and paging.

phase modulation (PM) One method of modulating a carrier signal. Angle modulation in which the phase angle of a carrier is caused to depart from its reference value by an amount proportional to the instantaneous value of the modulating signal. Other types of modulation are amplitude and frequency modulation.

phase-shift keying (PSK) A digital signal modulation technique in which the phase of the carrier is varied to represent characters, such as bits or quaternary digits. For example, when encoding bits, the phase shift could be 0° for encoding a "0," and 180° for encoding a "1," or the phase shift could be –90° for "0" and +90° for a "1," thus making the representations for "0" and "1" a total of 180° apart. In PSK systems designed so that the carrier can assume only two different phase angles, each change of phase carries one bit of information, *i.e.,* the bit rate equals the modulation rate. If the number of recognizable phase angles is increased to 4, then 2 bits of information can be encoded into each signal element; likewise, 8 phase angles can encode 3 bits in each signal element. Also called *biphase modulation, phase-shift signaling.*

physical topology The physical configuration, *i.e.,* interconnection, of network elements, such as cable paths, switches, and concentrators. Physical topology is in contrast to logical topology. For example, a logical loop may consist of a physical star configuration, or a physical loop.

ping Abbreviation for *packet Internet groper*. In TCP/IP, a protocol function that tests the ability of a computer to communicate with a remote computer by sending a query and receiving a confirmation response.

plant All the facilities and equipment used to provide telecommunications services. Plant is usually characterized as *outside plant* or *inside plant*. Outside plant, for example, includes all poles, repeaters and unoccupied buildings housing them, ducts, and cables—including the "inside" portion of inter-facility cables outward from the main distributing frame (MDF) in a central office or switching center. Inside plant includes the MDF and all equipment and facilities within a central office or switching center.

point of interface (POI) In a telecommunications system, the physical interface between the local access and transport area (LATA) access and inter-LATA functions. The POI is used to establish the technical interface, the test points, and the points of operational responsibility. Also called **interface point.**

point of presence (POP) A physical place within a local access and transport area (LATA) at which a long-distance carrier establishes itself for the purpose of obtaining LATA access and to which the local exchange carrier provides access services. This gives the long-distance carrier access to the local phone company's subscribers for originating and terminating long-distance calls.

point-to-multipoint transmission Communications from one origination point to more than one destination, such as TV or radio broadcasts.

point-to-point transmission Communications between two designated stations only.

PON See **passive optical network**.

POP See **point of presence**.

port On computers, telephones, and related devices, the physical place where the device connects with a cable to another device or to a network. On your telephone, for example, the port is the plug where the telephone line from the wall (and the public network) attaches to your phone. On your computer, there are ports to plug in the printer cable, a modem, and other peripheral devices.

POTS Abbreviation for *plain old telephone service*. The basic analog voice service consisting of standard telephones, telephone lines, and access to the public switched network. The term is typically used to distinguish the analog phone network from newer, digital services.

predictive dialing An automated method used by outbound call centers that dials a list of customers or prospects from a computer database and passes them to the operator only after a live person answers the telephone. In this way the operator does not have to spend time dialing the telephone and dealing with answering machines and busy signals. When you answer the telephone and there is a slight delay until the other party comes on the line, usually a telemarketer, he or she is probably using such a system.

premium billing services Billing and collection services provided by the telephone companies to information providers (IPs) or service bureaus for their pay-per-call information programs. Premium billing usually involves both the LEC and the IXC for national 900 number pay-per-call programs, with the LEC serving as the IXC's agent in collecting from the end customer in the monthly phone bill.

prepaid debit card See **prepaid phone card**.

prepaid phone card A card with a predetermined number of minutes or message units, normally used for long-distance calling. The card is tied to a prepaid phone card *platform*, which is essentially a computer switch that accepts the inbound call from the cardholder (usually via a toll-free 800 number), routes the outbound call to the desired party, keeps track of the message units consumed and remaining, and alerts the caller at certain intervals as the message units are depleted. Some prepaid phone cards also have a replenishment feature, where the cardholder can call into the system and buy additional message units with a credit card, using an IVR system to input all the data and to complete the transaction.

prepaid phone service Local telephone service for the credit-challenged who cannot or will not get regular phone service. The bill must be paid in cash before the service starts for the month, and it includes only local, toll-free, and 911 calls. Prepaid phone service providers usually market prepaid phone cards to their subscribers for long-distance calling.

primary rate interface (PRI) An integrated services digital network (ISDN) interface standard (a) that is designated in North America as having 24 (23B+1D) channels, (b) in which all circuit-switched B channels operate at 64 Kbps, and (c) in which the D channel also operates at 64 Kbps. The PRI combination of channels results in a digital signal 1 (T1) interface at the network boundary. See **integrated digital services network (ISDN)** for a detailed explanation.

priority ringing A service feature where you program in a list of phone numbers that will ring with a distinctive ring. In some areas, you can have only 10 numbers on this list. Other areas allow more numbers. This service is also called *priority call* and *call selector*. Priority ringing does not work if the person calling you is using a long-distance carrier. That is because the long-distance call does not carry the signalling in a manner that allows the local switch to read it. Priority ringing does not work with business lines that are part of a PBX or Centrex system. That is because the number that is sent to the central office computer is the company's main number, not the number of the extension dialing out.

private branch exchange (PBX) A private telephone switch usually owned by an organization and located on its premises. A PBX is used to direct calls among internal lines and between internal lines and outside lines, through the public telephone network. A PBX can be automatic (Private Automatic Branch Exchange—PABX) or manual (Private Manual Branch Exchange—PMBX). A PABX routes calls based on the number dialed and does not require an attendant to switch the call. A PMBX requires manual assistance. The primary benefit of installing a PBX is eliminating the need for each person within the organization to have an outside line—one that is connected to the telephone company's network. The PBX allows internal users to share outside lines. A PBX system usually includes a computer programmed to manage call switching, public network trunk lines, and in some cases, a switchboard for an operator. Digital PBXs, which convert signals to analog for calls on the local telephone network, have replaced most analog PBX systems.

private line In the telephone industry usage, a service that involves dedicated circuits, private switching arrangements, and/or predefined transmission paths, whether virtual or physical, which provide communications between specific locations. Among subscribers to the public switched telephone network(s), the term *private line* is often used to mean a one-party switched access line.

private telecommunications services Non-common-carrier telecommunications services, including private line, virtual private line, and private switched network services.

protocol Conventions or rules governing communications. Protocols exist for network architectures, communications technologies, and physical devices. Adherence to protocols allows networks to interconnect and users to exchange information, for example; it also ensures compatibility between devices made by different manufacturers. Protocols become *standards* when an official body, such as the Institute of Electrical and Electronic Engineers (IEEE), the International Standards Organization (ISO), or the International Telecommunications Union-Telecommunications Services Sector (ITU-T) signs off on them. This can often be a long, painstaking process. Lengthy delays frequently lead to the development of *de facto* standards that service providers and software/hardware equipment manufacturers agree to adhere to in order to bring new products and services to market more quickly.

PSTN See **public switched telephone network.**

PTT Abbreviation for *postal, telegraph, and telephone* (organization or administration). In countries having nationalized telephone and telegraph services, this is the generic term for the organization, usually a government department, which acts as its nation's common carrier.

public data network (PDN) A network established and operated by a telecommunications administration, or a recognized private operating agency, for the specific purpose of providing data transmission services for the public.

public land mobile network (PLMN) A network that is established and operated by an administration or by a recognized operating agency (ROA) for the specific purpose of providing land mobile telecommunications services to the public. A PLMN may be considered an extension of a fixed network, *e.g.* the Public Switched Telephone Network (PSTN) or as an integral part of the PSTN.

public switched network (PSN) Any common carrier network that provides circuit switching among public users. The term is usually applied to public switched telephone networks, but it could be applied more generally to other switched networks, *e.g.*, packet-switched public data networks.

public switched telephone network (PSTN) The global, landline telephone network accessible to everyone with an active connection who pays the required access fee.

public utilities commission (PUC) In the United States, a state regulatory body charged with regulating intrastate utilities, including telecommunications systems. In some states this regulatory function is performed by public service commissions or state corporation commissions.

PUC See **public utilities commission**.

pulse code modulation (PCM) A technology that allows an analog signal, such as voice, video, or combinations of voice and video, to be depicted in a digital format. For example, this is how the analog sound of a voice signal is *digitized* for transmission over a digital data network. The word *modulation*, in this context, means the method used to transmit data. A PCM device measures the analog signal at regular, very short intervals. Using the amplitude of the analog signal at each sampling, the transmitting PCM device represents the analog signal as a series of numbers. The receiver is equipped with a pulse code demodulator, which translates the numeric, or digital, transmission back into analog form.

Q

quadrature amplitude modulation (QAM) A modulation technique for transmitting digital signals that combines both phase and amplitude modulation. QAM employs two amplitudes and shifts the phase in one-quarter increments for each, resulting in eight possible combinations (bit values).

quality of service (QoS) A measure or standard of the level of service provided by the common carrier. Voice clarity, delay in receiving dial tone, or data transmission error rates are often quantified by public utility commissions (PUCs) in the their respective states. QoS over the established public switched telephone network (PSTN) is very high. The issue is with new data networks, such as packet-switched Internet protocol (IP) networks, where high QoS standards have yet to be established or widely implemented.

R

raceway Within a building, an enclosure, or a channel, used to contain and protect wires, cables, or bus bars.

radio common carrier (RCC) A common carrier engaged in the provision of Public Mobile Service, which is not also in the business of providing landline local exchange telephone service. These carriers were formerly called *miscellaneous common carriers.*

radio frequency (RF) Any frequency within the electromagnetic spectrum normally associated with radio wave propagation, between approximately 10 kHz and 300 MHz.

rate adaptive digital subscriber line (RADSL) A high-speed data transmission technology over twisted-pair copper wires that can detect the condition of the local loop and adapt the transmission rate so that it does not exceed the performance capabilities at any given instant. The total source-to-sink distance of the transmission and noise interference on the local loop can degrade the optimum data rate. See also **ADSL** and **DSL**.

RBOC See **Regional Bell Operating Company.**

rebiller See **reseller.**

reciprocal compensation A compensation arrangement between two carriers in which each of the two carriers receives compensation from the other carrier for the transport and termination on each carrier's network facilities of local telecommunications traffic that originates on the network facilities of the other carrier.

recognized operating agency (ROA) Any operating agency, or company, as defined in the ITU Convention (Geneva, 1992), that operates a public broadcasting service and is authorized to establish and operate a telecommunication service on its territory. Formerly called *recognized private operating agency (RPOA*).

refile The routing of international telephone traffic through an intermediate country in order to avoid the high **settlement rates** of the destination country. The intermediate country will have negotiated lower settlement rates with the destination country, so it ends up being less expensive for the carrier in the originating country to send traffic through the intermediate country, even when *those* settlement rates (charged by the intermediate country) are added to what the intermediate country charges to serve as the go-between to the destination country.

regional Bell operating company (RBOC) One of the seven holding companies formed by the divestiture of the American Telephone and Telegraph Company of its local Bell System operating companies, and to which one or more of the Bell System local telephone companies were assigned.

regional center See **office classification.**

registered jack (RJ) A series of modular telephone jacks described in the *Code of Federal Regulations*, Title 47, part 68. See also **RJ-11, RJ-14** and **RJ-45**.

repeat dialing A service feature that redials the last busy number you called and keeps trying until the line is free (for up to 30 minutes). If you dialed a number and reached a busy signal, you press a code. When the line is free, your phone will announce the call with a distinctive ring. This won't tie up your phone line like a redialer on your phone would. You can still make and receive calls. When the system detects that the desired party's line is clear, your phone will ring you. Also called *continuous redial.*

repeater **1.** An analog device that amplifies an input signal regardless of its nature, *i.e.,* analog or digital. **2.** A digital device that amplifies, reshapes, retimes, or performs a combination of any of these functions on a digital input signal for retransmission. The term *repeater* originated with telegraphy and referred to an electromechanical device used to regenerate telegraph signals. Use of the term has continued in telephony and data communications. In current usage it usually refers to a device that amplifies digital signals, while an **amplifier** handles analog signals.

resellers Companies that purchase wholesale telephone services, such as long-distance, from facilities-based carriers and resell the services under their own names/brands. These companies range in size from small operations to billion-dollar companies, many with annual revenues exceeding $100 million. Although these companies started by selling long-distance service, many are now expanding into local service, wireless resale, paging, Internet access, calling cards, and other telecommunications services. Some of these companies may own their own switching systems and other network facilities (often referred to as *switched* resellers), but they still lease lines from other carriers to complete their coverage. Those that do not own any network facilities are called *switchless* resellers.

resource reservation protocol (RSVP) This standard allows for different priorities to be assigned to individual data transmissions according to their quality-of-service (QoS) rating. Instead of the egalitarian method normally used on IP networks, where all data is treated equally, RSVP can give higher priority to a videoconference, for example, and a lower priority to e-mail. Network resources are actually reserved, across networks with varying topologies, between the sender and receiver. This allows a high-priority, high-bandwidth transmission to get through quickly in one uninterrupted transmission path.

return call A service feature that automatically calls back the last number that called you, even though you don't know the number. If the number you're dialing is free, your call will be completed. If it is busy, this service will check the line for you for up to 30 minutes and let you know when the line is free by ringing you with a distinctive ring. Nice time saver. Missed calls can easily be returned. Also called *call cue, call return* and *last call return.* This service does not work if the person calling you was calling using a phone system such as a PBX or Centrex system.

ringdown In telephony, a method of signaling an operator in which telephone ringing current is sent over the line to operate a lamp and cause the drop of a self-locking relay. Ringdown (a) is used in manual operation, as distinguished from dialing, (b) uses a continuous or pulsed alternating-current (AC) signal transmitted over the line, and (c) may be used with or without a switchboard. The term *ringdown* originated in magneto telephone signaling, in which cranking the magneto in a telephone set would not only *ring* its bell but also cause a lever to fall *down* at the central office switchboard.

ringer equivalency number (REN) A number determined in accordance with the *Code of Federal Regulations,* Title 47, part 68, which represents the ringer loading effect on a phone line. A ringer equivalency number of 1 represents the loading effect of a single traditional telephone set ringing circuit. Modern telephone instruments may have a REN lower than 1. The total REN expresses the total loading effect of the subscriber's equipment on the central office ringing current generator. The service provider usually sets a limit, *e.g.,* 3, 4, or 5 (representing extension, or parallel-connected telephones), to the total REN on a subscriber's loop. The actual number of instruments across the loop may be greater than the service provider's REN limit, if their respective individual RENs are less than 1.

ring topology A wiring or network geometry that looks like a ring. Every node has exactly two branches connected to it, creating a closed loop. See also **network topology.**

RJ-11 Registered jack number 11. The standard modular one-line jack with one pair of wires (two conductors), for single-line telephones, fax machines, modems, or other accessories. See also **registered jack.**

RJ-14 Registered jack number 14. Identical to the RJ-11 jack, except that it has two pairs of wires (four conductors), and is capable of connecting to one or two lines. See **also registered jack.**

RJ-45 Registered jack number 45. A standard 8-pin connector for data transmission over a single telephone line using one pair of wires (two conductors). See **also registered jack.**

rotary dial A signaling mechanism—usually incorporated within a telephone set—that when rotated and released, generates direct-current (DC) pulses required for establishing a connection in a telephone system.

rotary hunting Hunting in which all the numbers in the hunt group are selected in a prescribed order. In modern electronic switching systems, the numbers in the hunt group are not necessarily selected in consecutive order.

router A device (or software on a computer) that connects multiple networks and manages the flow of data between them, such as finding the best route and establishing communications priorities. A router directs each data packet it receives to the next network, and another router, which in turn moves the packet closer to its final destination. Routers operate at the network layer, which is layer 3 of the OSI model. Routers are similar to bridges but perform more sophisticated functions. A router on the Internet, for example, contains an index of all of the potential circuits an individual packet could take to reach its destination. The router also maintains constant knowledge of the current status of each circuit in the network it is connected to so that it can determine which course will be fastest and most cost effective. A router can be likened to a knowledgeable traffic cop who knows all of the different roads that lead to a particular destination and sends a driver along the route that is shorter (to save time and gasoline) and has the lightest traffic at that time of day.

routing table A matrix associated with a network control protocol, which gives the hierarchy of link routing at each node. Routers read the addresses on data packets and then consult their routing table to figure out where to send the data.

RSVP See **resource reservation protocol**.

S

satellite communications A telecommunications service provided via one or more satellite relays and their associated uplinks and downlinks.

screen pop When a customer service representative, order taker, or any other person in a call center fields an incoming call, the calling party's telephone number is often identified through caller ID or automatic number identification (ANI), and then matched to the caller's record in the company's database. The caller's record is then automatically displayed (pops) on the computer screen, which may show name, address, order history, model number, sizes or other preferences, helping the representative be as efficient as possible in helping the caller.

SDH See **Synchronous Digital Hierarchy.**

SDSL See **single-line digital subscriber line** and **symmetrical digital subscriber line.**

select call forwarding A service feature that allows you to forward a list of calls to an alternate number, and to restore the line to normal operation at your discretion. You program the list yourself and can update it whenever you choose. The calling party is not aware that the call is being forwarded. Useful for screening calls. Also called *preferred call forwarding*.

selective call acceptance A service feature that allows you to specify a list of up to 32 phone numbers from which calls will be accepted. When you activate this feature, all other callers hear a polite announcement telling them that you're not accepting their calls.

selective ringing In a party line, ringing only the desired user instrument. Without selective ringing, all the instruments on the party line will ring at the same time, selection being made by the number of rings.

serial port (interface) Serial ports are interfaces (plugs) that transmit signals across point-to-point data links one bit at a time in a serial stream. Serial ports found on PCs are usually labeled COM1 to COM4 and are invariably connected to modems. Parallel ports, on the other hand, are often connected to printers and transmit eight bits at a time (one byte), resulting in higher data throughput rates.

server A network device that provides service to the network users by managing shared resources. The term is often used in the context of a client-server architecture for a local area network (LAN). Examples are a printer server, a fax server, and a file server.

service access code See **area code**.

service access point (SAP) **1.** A physical point at which a circuit may be accessed. **2.** In an Open Systems Interconnection (OSI) layer, a point at which a designated service may be obtained.

service bureau In telecommunications, a company that provides voice processing and audiotext equipment and services, as well as connection to telephone network facilities. These companies can offer a variety of communications services, such as fax-on-demand, fax broadcasting, international callback, prepaid phone cards, interactive voice response (IVR) programs, 900 services, automated order processing, and other services. Service bureaus offer an alternative to purchasing equipment and operating such services in-house.

service termination point The last point of service rendered by a commercial carrier under applicable tariffs. The service termination point is usually on the customer premises. The customer is responsible for equipment and operation from the service termination point to user end instruments. The service termination point usually corresponds to the demarcation point.

session In data communications, the period during which a discrete block of data, from beginning to end, is transmitted between nodes on a network.

Session Layer See **Open Systems Interconnection**.

settlement rate The monetary rate at which countries pay one another for sending telephone traffic to the other country. Mathematically, the settlement rate is half the **accounting rate**, which is the average of the call-termination charges between the two countries. For example, if the settlement rate is $1, and 100,000 calls from country "A" terminate in country "B," but only 20,000 calls come back from country "B" to country "A," country "A" pays country "B" $80,000 for the difference (100,000-20,000=80,000).

seven hundred (700) service A service access code for use by interexchange carriers (IXCs) to provide specialized services of their own choosing. Some carriers have used the code for follow-me-anywhere services. Others use it for setting up virtual private networks (VPNs). You can dial 1-700-555-4242 to find out what long-distance carrier has been assigned the line you are calling out on.

SHF Abbreviation for **super high frequency.** See **electromagnetic spectrum.**

shortwave In radio communications, pertaining to the band of frequencies approximately between 3 MHz and 30 MHz. *Shortwave* is not a term officially recognized by the international community.

sideband In amplitude modulation (AM), a band of frequencies higher than or lower than the carrier frequency, containing energy as a result of the modulation process. Amplitude modulation results in two sidebands. The frequencies above the carrier frequency constitute what is referred to as the *upper sideband;* those below the carrier frequency, constitute the *lower sideband.* In conventional AM transmission, both sidebands are present. Transmission in which one sideband is removed is called *single-sideband transmission.*

signal computing system architecture (SCSA) An open set of hardware and software standards introduced by Dialogic Corporation (Parsippany, NJ) that allow communication and computing devices to work together. In this way, computer telephony applications can be built using components from different manufacturers.

signal control point (SCP) A computer that stores customer information in databases that can be accessed by the network through advanced intelligent network (AIN) capabilities such as Signaling System 7 (SS7).

Signaling System 7 (SS7) A network signaling system that improves network efficiency and allows for the provision of advanced services. Signaling System 7 is a means by which elements of the telephone network, such as switches and nodes, exchange information. The SS7 architecture is an integral part of the developing Advanced Intelligent Network (AIN) and Integrated Services Digital Network (ISDN). On an SS7 network, which is a broadband, packet-switched network, information is conveyed in the form of messages. Signaling messages can include information such as: the telephone number a call is coming from, the number it is going to, and the trunk on which the call is being sent; alerts about network congestion on a particular circuit; or a service outage.

In the past, telephone signaling took place over the same circuit that carried the telephone call itself. This method of signaling is called *inband.* SS7 is an *out-of-band* signaling system, meaning the messages between network elements take place on a separate high-speed network. Out-of-band signaling is faster (so calls can be set up more quickly) and frees bandwidth on the network carrying conversations, faxes or other exchanges.

The SS7 network in North America is built with three components—signal switching points (SSPs), signal transfer points (STPs) and signal control points (SCPs). These are connected by signaling links. SSPs are switches that originate, terminate and switch calls. STPs are the SS7 network's packet switches, used to route signaling messages to their destination. SCPs are databases that contain information that enables telephone companies to offer advanced services such as call forwarding, call waiting, caller ID, call blocking, automatic callback, and others.

Simple Mail Transfer Protocol (SMTP) The Transmission Control Protocol/Internet Protocol (TCP/IP) standard protocol that establishes the rules governing electronic mail transmissions between computers and other digital devices. SMTP governs the server-to-server portion of an e-mail transmission; other protocols are used to access the messages from the server.

Simple Network Management Protocol (SNMP) The Transmission Control Protocol/Internet Protocol (TCP/IP) standard protocol that (a) is used to manage and control IP gateways and the networks to which they are attached, (b) uses IP directly, bypassing the masking effects of TCP error correction, (c) has direct access to IP datagrams on a network that may be operating abnormally, thus requiring management, (d) defines a set of variables that the gateway must store, and (e) specifies that all control operations on the gateway are a side effect of fetching or storing those data variables, *i.e.,* operations that are analogous to writing commands and reading status.

simplex circuit A circuit that provides transmission in one direction only. Contrast with a **duplex circuit** that allows two-way transmission.

single-line digital subscriber line (SDSL) This is a variation of **high speed digital subscriber line (HDSL)** that operates over a single pair of wires instead of the normal two pairs with HDSL, although the local loop distance is constrained to about 10,000 feet. Also called *HDSL2*. See also **digital subscriber line (DSL)** for a detailed description of the technology.

slamming An illegal practice whereby service providers, usually long-distance carriers, switch subscribers away from other service providers to themselves without the subscribers' knowledge or consent.

SLIP Acronym for **serial line Internet protocol.** A protocol that allows a computer to use the Internet protocol (IP) with a standard telephone line and a high-speed modem.

SMDR See **station message-detail recording.**

SMDS See **switched multimegabit data services.**

SNA See **Systems Network Architecture**.

SNMP See **Simple Network Management Protocol.**

SONET See **synchronous optical network.**

spam Slang for unsolicited e-mail. The electronic equivalent of junk mail.

specialized common carrier (SCC) A common carrier offering a limited type of service or serving a limited market.

speech recognition The capability of a computer, using specialized software, to recognize human speech. This technology is frequently used in voice-processing systems, allowing callers to speak their instructions instead of using the telephone keypad. Also called *voice recognition.*

speed dialing A service feature that allows you program a list of phone numbers that you can dial using a one- or two-digit code. Some phone companies allow six numbers only; others offer 8, 30 or 50 numbers. To use, simply press the code. Frankly, phone service supplied-speed calling is rarely cost-effective. If you want speedy dialing, get a phone with this feature or, better yet, an autodialer program on your PC.

SS7 See **Signaling System No. 7.**

star (topology or wiring) All nodes or devices are connected to a central point or hub, and all communication must pass through the hub or centralized device, because the peripheral nodes or devices are not connected to one another. PBXs are usually wired this way. See **network topology**.

station A telephone instrument. A term used by telephone equipment vendors.

station message-detail recording (SMDR) A record of all calls originated or received by a switching system. SMDRs are usually generated by a computer.

statistical mutiplexing Other multiplexing methods such as time division multiplexing (TDM) give each device on a network equal time to send data in a timed sequence, whether the device has data to send or not. Statistical multiplexing polls each device first, and gives it an open channel only when it has data to send out. This is more efficient with network resources.

streaming (audio/video) The transmission of digital audio or video files over a packet-switched network, such as the Internet, from a server where the files are stored to the end user who has requested the file. The transmission initially goes into a buffer in the recipient's computer and is then played from the buffer, eliminating problems of delayed or out-of-sequence packets and resulting in an uninterrupted data stream from the user's perspective.

subnet address In an Internet Protocol (IP) address, an extension that allows users in a network to use a single IP network address for multiple physical subnetworks. The IP address contains three parts: the network, the subnet, and host addresses. Inside the subnetwork, gateways and hosts divide the local portion of the IP address into a subnet address and a host address. Outside of the subnetwork, routing continues as usual by dividing the destination address into a network portion and a local portion.

subnetwork A collection of equipment and physical transmission media that forms an autonomous whole and that can be used to interconnect systems for purposes of communication. A LAN is limited (by technology) to a certain number of devices, and the use of multiple subnetworks can overcome this limitation, resulting in several interconnected LANs.

subscriber In a public switched telecommunications network, the ultimate user, *i.e.,* customer, of a communications service. Subscribers include individuals, activities, organizations, etc. Subscribers use end instruments, such as telephones, modems, facsimile machines, computers, and remote terminals, that are connected to a central office.

switch A computer on a telecommunications or data network that chooses the next destination on the network for a voice call or digital transmission, locates an open circuit to that destination, and sends the call or data there. A switch can also open or close a circuit at the beginning or end of a phone call.

switch-based reseller A company that purchases telephone services in large wholesale volume from facilities-based carriers at big discounts, while also owning one or more switches and, in some cases, transmission equipment.

switched 56 service This is a switched dial-up digital data link operating at 56 Kbps, designed for occasional or intermittent use only as necessary (as opposed to a dedicated leased line, for example). It is useful for videoconferences, LAN-to-LAN connections, high-speed bulk file transfer, distance learning, video surveillance, electronic publishing, video conferencing, shared library transfer, and image transfers (such as advertising layouts, real estate listings, and digitized fingerprints). Although this is an established, reliable service that has been around for some time, it is being replaced by faster ISDN and DSL services.

switched circuit In a communications network, a circuit that may be temporarily established at the request of one or more of the connected stations.

switched loop In telephony, a circuit that automatically releases a connection from a console or switchboard, once the connection has been made to the appropriate terminal. Loop buttons or jacks are used to answer incoming listed directory number calls, dial "0" internal calls, transfer requests, and intercepted calls. The attendant can handle only one call at a time. Synonym **released loop.**

switched multimegabit data services (SMDS) A broadband, packet-switched network service that allows organizations to connect their private local area networks (LANs) over the public network. SMDS eliminates the need for organizations to lease more expensive private lines to accomplish the same sort of LAN-to-LAN connections. Any SMDS customer can link its LAN to a carrier's public switched fiber-optic network to exchange information at T1 and T3 speeds (1.544 Mbps and 44.736 Mbps) with any other SMDS customer. SMDS can be used to create both intranets, for organizations with LANs in remote locations, and extranets, extending most of the benefits of a LAN over a much greater distance and between separate organizations. As is the case in other packet-switched networks, each distinct data packet is sent separately as the bandwidth is available. Each packet contains the addresses of both the sender and receiver. An association of SMDS carriers, users and equipment manufacturers, The SMDS Interest Group, works together to ensure global interoperability and to encourage the development of new SMDS services.

switched network **1.** A communications network, such as the public switched telephone network, in which any user may be connected to any other user through the use of message-, circuit-, or packet-switching and control devices. **2.** Any network providing switched communications service.

switching center In communications systems, a facility in which switches are used to interconnect communications circuits on a circuit-, message-, or packet-switching basis. Synonyms, in telephony, **central office, switching exchange,**

switching facility, or simply **switch.**

switchless reseller Companies that do not own a switching system or any other network facilities, but purchase telephone services in large wholesale volume from facilities-based carriers or larger resellers at big discounts. These companies then sell to subscribers under their own brand or name, performing all marketing, billing, and collection functions.

symmetrical digital subscriber line (SDSL) A nonstandard variation of **high speed digital subscriber line (HDSL)**. See **digital subscriber line (DSL)** for a detailed description of the technology.

synchronous data link control (SDLC) In a data network, a bit-oriented protocol for the control of synchronous transmission over data links.

Synchronous Digital Hierarchy (SDH) The worldwide fiber-optic telecommunication standard adapted by the International Telecommunications Union (ITU). SONET is the North American version of SDH. SDH allows telecommunications companies worldwide to interconnect with one another by using the same standards and protocols in their fiber-optic transmission systems. Most experts believe SDH/SONET will continue to provide the technology for high-speed network backbones. SDH standardizes transmission around the bit rate of 51.84 megabits per second, which is also called *STS-1*. Using the SDH, nodes on a fiber-optic network do not have to use the same synchronized clock to send or receive data. This is an advantage over other multiplexing technologies, which can require users to implement proprietary methods of synchronizing the network. Experts say SDH is more flexible and results in more reliable networks and better performance at lower operating and maintenance costs. Using SDH, synchronization is achieved by adding network management bits to the data traffic stream before it is multiplexed into one of the STS fixed-frame rates. SDH framing is done at a fundamental clock frequency of 8 KHz, or 125 microseconds. See also **synchronous optical network (SONET)**.

synchronous optical network (SONET) A standard for using fiber-optics for telecommunications transmissions. SONET is likely to become the basis for global telecommunications well into the next century. SONET is the standard adopted by the American National Standards Institute (ANSI) for the United States; the global equivalent standard is known as **Synchronous Digital Hierarchy** (SDH). SONET comprises a group of transmission rates for fiber-optic lines ranging from 51.84 Mbps to 13.22 Gbps, moving upward in multiples of 51.84 Mbps. (This is the "base rate" of SONET, the initial speed of SONET's Optical Carrier signal, which originates as a synchronous transport signal [STS]. Therefore, the transmission speed of 51.84 Mbps is known as *STS-1*.). By building to the SONET standard, transmission equipment built by various vendors can interface and work together. SONET is a high-bandwidth standard that will provide broadband optical fiber connections in a point-to-point configuration. SONET will support services based on ATM. SONET's advantages over the older, copper-based telephone network include increased network reliability, interconnection between products from different vendors, and a defined architecture that is flexible enough to support new applications and a range of transmission speeds. SONET transmission involves weaving together frames of data in a synchronous high-speed signal that forms STS-1, or a higher multiple of STS. SONET networks are usually laid out in a ring and are redundant, meaning more than one optical fiber can transmit signals across the same geographical area in the event one fiber fails. Many local networks use fiber-optic transmission lines, known as *SONET rings*. In addition to the enormous bandwidth they offer, allowing them to carry many calls and/or large amounts of data, SONET rings have separate channels that carry information about the performance of the network. Such information is known as *operations, administration, maintenance and provision* (OAM&P), for which SONET has a specific protocol. The OAM&P protocol allows carriers to manage the network, introduce new services, bill for their use, and to instantly reroute calls if a portion of the SONET network is broken.

synchronous transmission See **asynchronous transmission** for a full description of both transmission types.

Systems Network Architecture (SNA) IBM's architecture for computer networks. SNA embodies both the protocols (or rules) for communication among devices on an SNA network as well as the hardware and software that makes up the network itself. SNA preceded IBM's SAA (Systems Application Architecture) and has become a part of it. As more organizations have built networks incorporating products from different vendors and the Internet, with its *de facto* TCP/IP standard, IBM has developed methods to combine SNA with other network architectures.

T

tandem office A central office that serves local subscriber loops, and also is used as an intermediate switching point for traffic between central offices.

TAPI See **telephone application programming interface**.

tariff The published schedule of rates or charges for a specific unit of equipment, facility, or type of service, which are filed by a regulated telephone company with a state public utility commission or the Federal Communications Commission. The tariff describes services, equipment and pricing offered by the telephone company to all potential customers. As a **common carrier**, the telephone company must offer its services to the general public at the prices and conditions outlined in its tariffs.

T-carrier A system for carrying voice transmissions in a digital format. The first transmission rate, 1.544 Mbps, is in wide use today and is known as T-1 (or T1). T-carrier systems use pulse code modulation and time-division multiplexing to deliver multiple voice or data calls simultaneously. T-carrier systems, which were developed in the 1960s, initially used twisted-pair copper wire. Today, the transmission medium can be optical fiber, coaxial cable, digital microwave, and others. T-carrier links correspond to the digital service (DS) level hierarchy. (See **DS-0**, **DS-1**, etc.) T-carrier circuits range in capacity from 1.544 Mbps (T-1) to 274.176 Mbps (T-4), as outlined below:

> **T-1 (T1):** A digital transmission link with a capacity of 1.544 Mbps, that can be divided into 24 voice channels, each operating at 64 Kbps, by multiplexing. The "voice" channels can also be used to carry data. In fact, T-1 lines frequently are used to connect individual networks to the Internet. The entire 1.544 Mbps bandwidth or any portion thereof can be used as a single high-speed data channel (*i.e.*, 12 voice channels and one 768 Kbps data channel). Telephone companies lease T-1 lines (or parts of T-1 lines, called *fractional T-1s*) to customers requiring high-capacity lines. T-1 circuits frequently are used to link remote LANs.
>
> **T-2:** Four times the capacity of T-1 at 6.312 Mbps, capable of 96 voice conversations or data links.
>
> **T-3:** Twenty-eight T-1 lines, or 44.736 Mbps, capable of 672 voice conversations or data links, carried on fiber-optic cable.

T-4: 168 times the capacity of T-1 at 274.176 Mbps, capable of handling 4032 voice conversations or data links.

TCP See **Transmission Control Protocol.**

TCP/IP Abbreviation for **Transmission Control Protocol/Internet Protocol.** Two interrelated protocols that are part of the Internet protocol suite. The primary networking protocols, or system of communication, used on the Internet, and now on intranets and extranets. TCP/IP itself is a two-layer protocol that first divides a message or file into packets for transmission over the Internet. These packets are reassembled into the original message on the receiving end. This is the Transmission Control Protocol (TCP) part. Each packet comprising an entire message or file must be "stamped" with an address so it can be routed through the Internet (or an Internet-like network) to its final destination. Each packet also carries the source address, or the address of the sender. This addressing is controlled through the Internet protocol (IP). Over time, TCP/IP has become a shorthand expression that can be used to refer to the many different protocols used on the Internet, such as Simple Mail Transfer Protocol (SMTP), the Hypertext Transfer Protocol (HTTP) used on the World Wide Web, and the File Transfer Protocol (FTP). PCs that are equipped for Internet access include the TCP/IP program, and usually programs for the additional protocols mentioned above, so the user can send and receive information over the Internet.

TCP/IP governs point-to-point communication based on the client/server method of networking computers. In other words, Internet users (or clients) ask for information from computers (servers) on the Internet. TCP/IP is considered a connectionless protocol because, unlike the dedicated circuit used to support a telephone conversation, communications between two points require no initial setup and the packets being exchanged can take different routes through the network. Among the reasons TCP/IP is becoming so widely used is because it's such a flexible protocol. It can be used with a variety of other protocols and physical communication mediums. TCP/IP can be used on backbone networks connecting Ethernet LANs, for example, and can be used on T-1 lines and X.25 networks.

TCP/IP Suite The suite of interrelated protocols associated with Transmission Control Protocol/Internet Protocol. The TCP/IP Suite includes, but is not limited to, protocols such as TCP, IP, UDP, ICMP, FTP, and SMTP. Additional application and management protocols are sometimes considered part of the TCP/IP Suite. This includes protocols such as SNMP.

TDD See **Telecommunications Device for the Deaf.**

TDDRA See **Telephone Disclosure and Dispute Resolution Act**.

TDM See **time-division multiplexing.**

TDMA See **time-division multiple access.**

telco Short for *telephone company*. Usually refers to the local phone company, but with long-distance carriers getting into local service and vice versa, the term will eventually encompass all telecommunications service providers.

telecard See **prepaid phone card**.

telecom Short for *telecommunication(s)*.

telecommunication Any transmission, emission, or reception of signs, signals, writings, images, sounds, or information of any nature by wire, radio, visual, optical, or other electromagnetic systems.

Telecommunications Act of 1996 The federal law enacted for the purpose of deregulating telecommunications, by allowing long-distance carriers into the local market, and allowing the local phone companies into long-distance markets (after certain conditions are met). Further goals of the Act are to encourage competition in order to keep prices low and to foster the development of new technology. Implementation has been difficult and uneven.

Telecommunications Device for the Deaf (TDD) A machine that uses typed input and output, usually with a visual text display, to enable individuals with hearing or speech impairments to communicate over a telecommunications network.

telecommunications management network (TMN) A network that interfaces with a telecommunications network at several points in order to receive information from, and to control the operation of, the telecommunications network. A TMN may use parts of the managed telecommunications network to provide for the TMN communications.

Telecommunication Relay Service (TRS) A relay service for the benefit of hearing and speech impaired people. A communications assistant (CA) facilitates the interaction between the impaired person, who might use a Telecommunications Device for the Deaf (TDD), for example, and another person, who may be impaired or not. The goal is for impaired people to be able to use the telephone in a manner functionally equivalent to non-impaired people.

telecommuting The act of performing one's work from a location other than the office through one or more telecommunications links to the main office. For example, employees can telecommute from home part of the time or full time, often using a high-speed ISDN line to the office. High-speed lines, fax machines, teleconferencing, document conferencing, and other remote-working technologies are making it progressively easier for knowledge-based workers to accomplish their work from virtually anywhere. Also called *teleworking*.

teleconference The live exchange of information among persons and machines remote from one another but linked by a telecommunications system. The term generally refers to voice conferencing. *Videoconferencing* refers to the transmission of video images along with voice and data.

telephone application programming interface (TAPI) An application programming interface (API) developed by Microsoft Corp. specifically to support telephones and accessories such that these communications devices and the related application programs can function together under the Windows operating systems.

Telephone Disclosure and Dispute Resolution Act (TDDRA) The federal law that governs the pay-per-call industry, such as 900 number services. It mandates disclosure requirements and outlines the responsibilities of service providers while providing relief for consumers for disputed charges.

telephone exchange See **central office.**

telephony The branch of science devoted to the transmission, reception, and reproduction of sounds, such as speech and tones. Transmission may be via various media, such as wire, optical fibers, or radio. Analog representations of sounds may be digitized, transmitted, and, on reception, converted back to analog form. *Telephony* originally entailed only the transmission of voice and voice-frequency data. Currently, it includes new services, such as the transmission of graphics information.

Telephony Server Application Programming Interface (TSAPI) An application programming interface (API) developed by AT&T to establish standards for call routing, device monitoring/querying, database transactions, and other network operations, allowing devices and applications from various vendors to interact seamlessly.

teletypewriter (TTY) A printing telegraph instrument that has a signal-actuated mechanism for automatically printing received messages. A TTY may have a keyboard similar to that of a typewriter for sending messages. Radio circuits carrying TTY traffic are called *RTTY circuits* or *RATT circuits*.

television (TV) A form of telecommunication for the transmission of transient images of fixed or moving objects. The picture signal is usually accompanied by the sound signal. In North America, TV signals are generated, transmitted, received, and displayed in accordance with the NTSC standard.

Telnet The TCP/IP standard network virtual terminal protocol that is used for remote terminal connection service and that also allows a user at one site to interact with systems at other sites as if that user terminal were directly connected to computers at those sites.

terahertz (THz) A unit denoting one trillion (10^{12}) hertz.

terminal A device capable of sending, receiving, or sending and receiving information over a communications channel.

Terminal Access Controller (TAC) A host computer that accepts terminal connections, usually from dial-up lines, and that allows the user to invoke Internet remote log-on procedures, such as Telnet.

terminal adapter An interfacing device that allows connection of a non-ISDN terminal (*i.e.*, telephone) with an ISDN network. Typically, a terminal adapter will support standard RJ-11 telephone connection plugs for voice and RS-232C, V.35 and RS-449 interfaces for data.

terminal equipment **1.** Communications equipment at either end of a communications link, used to permit the stations involved to accomplish the mission for which the link was established. **2.** Telephone and telegraph switchboards and other centrally located equipment at which communications circuits are terminated.

three-way calling A switching system service feature that permits users to add a third party at a different number during a call, without the assistance of an attendant. How it works: Dial the number of the first party you wish to link, tell them what you're up to, then press hookflash quickly. This puts the first party on hold and gives you a second dial tone. Dial the second number, tell them that you're going to connect the first party into the conversation, and press hookflash again. This puts you all together. You can also use this service to place a caller on hold, and place another call while holding the first. This lets you use it as though you had an extra outgoing line.

THz See **terahertz**.

tie trunk A telephone line that directly connects two private branch exchanges (PBXs).

time-division multiple access (TDMA) In digital cellular communications, a method of increasing the traffic a cellular channel can carry by dividing it into distinct slots, based on units of time. Other methods of increasing the capacity of a cellular channel include FDMA (see **frequency division multiple access**) and CDMA (see **code division multiple access**). TDMA, by allocating unique time slots to individual digital cellular users, allows a single radio-frequency channel to support multiple users. In TDMA, each channel is divided into six time slots. Since each signal requires two time slots, TDMA allows three signals to be multiplexed over the same channel. The result is a three-to-one gain in capacity. TDMA systems can support voice, data, message and fax services. TDMA works because the audio signal has been transformed into a digital signal that consists of discrete information packets that require just milliseconds to transmit. As a result, TDMA can allocate a single frequency channel for a short time, then switch to another channel, then to another, and back again. By bouncing between the three channels very, very quickly, TDMA allows all three conversations (or faxes or messages, etc.) to be transmitted virtually simultaneously. Early implementations of TDMA tripled the capacity of cellular transmissions by dividing a 30-kHz channel into three channels. There are now systems in place that transmit six times original capacity. Some experts estimate that future digital cellular capacity, using TDMA systems, could approach 40 times analog cellular capacity.

time-division multiplexing (TDM) Transmitting several digital signals on a single channel—either a telephone line, a satellite system or a microwave system—by sending each signal at a slightly different time. Each signal is assigned a time slot, and the time division is so small that each transmission appears to be constant and the multiple signals seem simultaneous. Each separate transmitting device sharing a TDM line is assigned its time slot based on the share of bandwidth it needs to deliver data. Since TDM awards a time slot to each channel sharing the line, whether it has information to transmit or not, a standard TDM system wastes bandwidth unless traffic on all channels is constant. Where traffic is intermittent, *statistical TDM* (STDM) can be used. Rather than assigning specific time slots to each channel, as TDM does, STDM allows each channel to transmit whenever a time slot is available. In essence, the channels "compete" for a time slot. An STDM system has the capacity for short-term memory, so when traffic on an STDM line is heavy, it can store data until a time slot is available. See also **statistical multiplexing**.

T-interface For basic rate access in an Integrated Services Digital Network (ISDN) environment, a user-to-network interface reference point that (a) is characterized by a four-wire, 144-Kbps (2B+D) user rate, (b) accommodates the link access and transport layer function in the ISDN architecture, (c) is located at the user premises, (d) is distance-sensitive to the servicing network terminating equipment, and (e) functions in a manner analogous to that of the Channel Service Units (CSUs) and the Data Service Units (DSUs).

tip & ring A telephony term that refers to the physical components of the old electrical plugs used in manual switching offices. The tip and the ring portion of the plug were connected to two different conductors (wires). The term has since come to refer to the two wires connecting a single-line telephone and creating the local loop connection between the subscriber and the central office. The tip wire connects the transmitter and the ring wire goes to the receiver of the telephone.

token In certain local-area-network protocols, a group of bits that serves as a symbol of authority, is passed among data stations and is used to indicate the station that is temporarily in control of the transmission medium.

token-bus network A bus network in which a token-passing procedure is used.

token ring adapter A network interface card (NIC) designed to attach a client workstation to a token ring computer network and operate as a token-passing interface.

token ring Along with Ethernet, one of the two most prevalent transmission technologies for packet-switched LANs, initially developed by IBM and then standardized by the Institute of Electrical and Electronic Engineers (IEEE). The name *token ring* is derived from two characteristics of the networks built using this protocol. The word *token* refers to the traffic-control method and the word *ring* refers to the configuration, or shape, of the network. Token ring networks are designed in a ring shape—sort of. Actually, token ring networks can be wired in a circle or a star shape. A popular wiring technique for token ring networks looks more like a collapsed ring, or star, since PCs and other devices on the network usually are laid out in a circle but each is physically connected to a central device called a *multi-station access unit (MAU)*. The MAU, a small box with eight or 16 connectors, is wired to create a circle that links the devices on the network. If there is a problem with a PC or section of wire on a token ring LAN, a MAU can recreate the ring by skipping that portion of the network. MAUs can be linked to each other to accommodate more devices on the LAN.

Token ring networks are designed so that only the device that has a special *token* can send data over the network. The token is a special data packet. By limiting transmission to the PC or workstation holding the token, collisions of packets sent by different devices can be avoided. Packets on a token ring network travel from one device to another, around the ring, until they reach their destination. The receiving device copies the packet and sends the original packet on around the ring. When the sending device receives the packet it originally sent back again, it pulls it off the network and sends a free token to the next device. If that device has data packets to send, it removes the free token and sends its packet. If not, it passes the free token to the next device. Token ring networks operate at 4 Mbps or 16 Mbps and can link up to about 100 PCs, although as many as 256 could be accommodated. Separate token ring LANs can be connected using bridges. Token ring networks have roughly twice the capacity of Ethernet, while FDDI has five to six times more capacity than token ring.

token-ring network See **network topology** and **token ring**.

toll call See **long-distance call.**

T1 through **5 (carrier)** See **T-carrier.**

traffic **1.** The information moved over a communication channel. **2.** A quantitative measurement of the total messages and their length, expressed in CCS or other units, during a specified period of time.

transceiver A single device that performs both transmitting and receiving functions.

transfer call A feature that gives you the ability to transfer a call from your phone to another phone. This saves you from having to ask the caller to hang up and dial the other number.

transmission **1.** The dispatching, for reception elsewhere, of a signal, message, or other form of information. **2.** The propagation of a signal, message, or other form of information by any means, such as by telegraph, telephone, radio, television, or facsimile via any medium, such as wire, coaxial cable, microwave, optical fiber, or radio frequency. **3.** In communications systems, a series of data units, such as blocks, messages, or frames.

Transmission Control Protocol (TCP) In the Internet protocol (IP) suite, a standard, connection-oriented, full-duplex, host-to-host protocol used over packet-switched computer communications networks. TCP controls how the data packets are created at the origin and reassembled at the destination.

transmission rate See **data rate.**

Transport Layer See **Open Systems Interconnection (OSI).**

tremendously high frequency (THF) Frequencies from 300 GHz to 3000 GHz.

trunk A communication line connecting two switching systems, such as a line connecting a central office and a PBX, or two central offices. Often used to refer to the communication line(s) with the most bandwidth or capacity on the network, the main line. A *tie trunk* connects PBXs, while *central office trunks* connect a PBX to the switching system at the central office.

trunk group Two or more trunks of the same type between two given points.

trunk group multiplexer (TGM) A time-division multiplexer that combines individual digital trunk groups into a higher rate bit stream for transmission over wideband digital communications links.

TSAPI See **Telephony Server Application Programming Interface**.

TTY See **teletypewriter.**

TTY/TDD A unique telecommunication device for the deaf, using TTY principles.

twisted-pair cable A cable made up of one or more separately insulated twisted-wire pairs, usually made of copper. This is the most common transmission media, used in LANs and between central offices and end-users/subscribers in the local loop. The thicker the wire, the better the transmission capabilities, and twisted-pair cable is designated Category 3 (voice-grade, data rates up to 16 Mbps) through Category 5 (data rates up to 100 Mbps).

U

UHF See **ultra high frequency.**

ULF See **ultra low frequency.**

ultra high frequency (UHF) Frequencies from 300 MHz to 3000 MHz.

ultra low frequency (ULF) Frequencies from 300 Hz to 3000 Hz.

unbundled network element (UNE) An individual, stand-alone service normally provided by a local phone company (LEC) that can be separated from other services, such as operator services or directory information services. See also **unbundling**.

unbundling The process of separating individual tariffed offerings and services from other tariffed basic service offerings. Incumbent local exchange carriers (ILECs) are required to unbundle their service offerings so that competitive local exchange carriers (CLECs) can purchase them wholesale for resale to their customers.

unified messaging The merging of all messaging, including voice mail, fax, and e-mail, into one easy-to-manage system. The integration is usually accomplished on a local area network (LAN) that connects the telephone system, voice mail system, Internet access and fax servers. The user sees all messages from diverse sources on one screen, and is better able to manage and respond to the messages. Also called *integrated messaging*.

universal personal telecommunications number See **UPT number.**

Universal Personal Telecommunications (UPT) service A telecommunications service that provides personal mobility and service profile management. UPT service involves the network capability of identifying uniquely a UPT user by means of a UPT number. UPT and PCS are sometimes mistakenly assumed to be the same service concept. UPT allows complete personal mobility across multiple networks and service providers. PCS may use UPT concepts to improve subscriber mobility in allowing roaming to different service providers, but UPT and PCS are not the same service concept.

Universal Resource Locator (URL) An address identifying sites on the World

Wide Web. Also called a Web site address. Our address is *http://www.aegisbooks.com*.

universal service The concept of making basic local telephone service (and, in some cases, certain other telecommunications and information services) available at an affordable price to all people within a country or specified area, including expensive-to-reach rural areas.

unshielded twisted pair (UTP) A cable made up of one or more separately insulated twisted-wire pairs, usually made of copper. This is the most common transmission media, used in LANs and between central offices and end-users/subscribers in the local loop. It is unshielded because it is not encased in a metallic sheath. See also **twisted-pair cable**.

uplink (U/L) **1.** The portion of a communications link used for the transmission of signals from an Earth terminal to a satellite or to an airborne platform. An uplink is the converse of a downlink. **2.** Pertaining to data transmission from a data station to the head end.

upstream **1.** The direction opposite the data flow. **2.** With respect to the flow of data in a communications path: at a specified point, the direction toward which data are received earlier than at the specified point.

UPT See **Universal Personal Telecommunications service**.

UPT access code In universal personal telecommunications service, the code that UPT users may need to dial, when using some terminals and networks, to enter the UPT environment before executing any UPT procedures.

UPT number In universal personal telecommunications, the number that uniquely identifies a UPT user and that is used to place a call to, or to forward a call to, that user. A user may have multiple UPT numbers, *e.g.,* a business UPT number for business calls and a private UPT number for private calls.

URL See **Universal Resource Locator**.

User Datagram Protocol (UDP) In the Internet protocol (IP) suite, a standard, low-overhead, connectionless, host-to-host protocol that is used over packet-switched computer communications networks, and that allows an application program on one computer to send a datagram to an application program on another computer. The main difference between UDP and TCP is that UDP provides connectionless service, whereas TCP does not.

UTP See **unshielded twisted pair**.

V

value-added carrier A company that sells the services of a **value-added network**.

value-added network (VAN) A network using the communication services of other commercial carriers, using hardware and software that permit enhanced telecommunication services to be offered. For example, an entity may purchase network transmission and switching services from established carriers and then add other services to the network, such as real-time stock quotes or voice mail services. Even an X.25 packet-switched network is considered a VAN because the network adds error checking and other rcliability services to the basic data transport service.

VAN See **value-added network**.

VC See **virtual circuit.**

very-high-data-rate digital subscriber line (VDSL) A high-speed version of **ADSL**, capable of data rates up to 51.84 Mbps. See **digital subscriber line (DSL)** for a detailed description of the technology.

very high frequency (VHF) Frequencies from 30 MHz to 300 MHz.

very low frequency (VLF) Frequencies from 3 kHz to 30 kHz.

VHF See **very high frequency.**

videoconference A teleconference that includes video communications, consisting of a two-way electronic communications system that permits two or more persons in different locations to engage in the equivalent of face-to-face audio and video communications. Videoconferences may be conducted as if all of the participants are in the same room. The PC-based desktop models of this equipment consist of a little video camera mounted on the monitor, a video card, speakers, and software to make it all work. The software may also support *document conferencing*, where both parties can work on the same document simultaneously, each seeing what the other changes on the document. This technology works best on higher-speed connections such as ISDN, because the performance on regular analog phone lines is fuzzy, jerky, and slow.

virtual circuit (VC) A communications arrangement in which data from a source user may be passed to a destination user over various real circuit configurations during a single period of communication. In data networks using packet-switching, for example, each individual data packet may actually take a different path from source to destination. A virtual circuit is a logical circuit, not a physical point-to-point connection. Virtual circuits are generally set up on a per-call basis and are disconnected when the call is terminated; however, a permanent virtual circuit can be established as an option to provide a dedicated link between two facilities. Also called *logical circuit* or *logical route.*

virtual network A network that provides virtual circuits and that is established by using the facilities of a real network. See also **virtual circuit** and **virtual private network.**

virtual private network (VPN) The use of the public telecommunications network, along with encryption and other security measures, to create a network that functions like a private data network but at a lower cost. True private networks are systems of communications lines that are owned or leased by a single company, and can only be used by that company. Leasing lines to create a private network is quite expensive, so service providers developed an alternative: the virtual private network. A virtual private network uses encryption to provide security over a group of communication lines, like the Internet, that is used by many people and organizations. The Point-to-Point Tunneling Protocol (PPTP) has been proposed by several leading computer and communications hardware and software vendors as the standard encryption/security protocol for virtual private networks.

virtual terminal (VT) In open systems, an application service that (a) allows host terminals on a multi-user network to interact with other hosts regardless of terminal type and characteristics, (b) allows remote log-on by local-area-network managers for the purpose of management, (c) allows users to access information from another host processor for transaction processing, and (d) serves as a backup facility.

VLF See **very low frequency.**

voice dialing A service feature that lets you program your own personal telephone directory by saying a name in any language and keying in the corresponding phone number. Henceforth, you just pick up the phone, say the name ("broker" or "mom") and the call is automatically dialed. No special equipment is needed; this service works with rotary as well as touch-tone phones.

voice frequency (VF) Pertaining to those frequencies within that part of the audio range that is used for the transmission of speech. In telephony, the usable voice-frequency band ranges from approximately 300 Hz to 3400 Hz. The bandwidth allocated for a single voice-frequency transmission channel is usually 4 KHz.

voice grade In the public regulated telecommunications services, a service grade that is described in part 68, Title 47 of the *Code of Federal Regulations [CFR].* Voice-grade service does not imply any specific signaling or supervisory scheme.

voice mail system A computerized system that records, stores and retrieves voice messages. You can program the system (voice mailboxes) to forward messages, leave messages for inbound callers, add comments and deliver messages to you, etc. It is essentially a sophisticated answering machine for a large business with multiple phone lines (probably with a PBX), or it can be a network-based service provided by the phone company.

voice over IP (VoIP) The technology that turns voice conversations into data packets and sends them out over a packet-switched Internet protocol (IP) network. See **IP telephony** for a detailed explanation.

voice processing This is the general term encompassing the use of the telephone to communicate with a computer by way of the touch-tone keypad and synthesized voice response. Either synthesized or recorded human voice can be stored on a computer's hard drive for later retrieval by use of the telephone keypad in responding to the voice-processing system's menu options. Audiotex, speech recognition, and interactive voice response (IVR) are subclassifications of voice processing.

voice recognition The ability of a computer to recognize human speech and the spoken word. Voice recognition is used in some voice-processing systems, allowing callers to state their preferences instead of using the telephone keypad to make selections. Also known as *speech recognition.*

voice response unit (VRU) This is the building block of any voice-processing system, essentially a voice computer. Instead of a computer keyboard for entering information (commands), a VRU uses remote touch-tone telephones. See also **interactive voice response** and **voice processing**.

VoIP See **voice over IP**.

V-series Recommendations Sets of telecommunications protocols and interfaces defined by CCITT (now ITU-T) Recommendations. Some of the more common V.-series Recommendations are:

V.21: A CCITT Recommendation for modem communications over standard commercially available lines at 300 bps. This protocol is generally not used in the United States.

V.22bis: A CCITT Recommendation for modem communications over standard commercially available voice-grade channels at 2,400 bps and below.

V.32: A CCITT Recommendation for modem communications over standard commercially available voice-grade channels at 9.6 Kbps and below.

V.32bis: A CCITT Recommendation for modem communication over standard commercially available voice-grade channels at 14.4 Kbps and below.

V.34: An ITU-T Recommendation for modem communication over standard commercially available voice-grade channels at 28.8 Kbps and below.

V.42: A CCITT Recommendation for error correction on modem communications.

V.42bis: A CCITT Recommendation for data compression on a modem circuit.

V.FAST: A new CCITT Recommendation for high-speed modems.

W

WAN See **wide area network.**

WATS See **Wide Area Telephone Service.**

wavelength The distance between points of corresponding phase of two consecutive cycles of a wave. For example, the distance between the peaks on a sine wave.

wavelength-division multiplexing (WDM) Provides significant increases in network capacity by allowing separate optical channels using different wavelengths of laser light to share the same fiber-optic line. By using different colors of light—each corresponding to a different wavelength—for each separate channel of information, the data streams can be kept separate until they reach their destination on the network, where they are separated by diffraction grating that identifies each color. Using WDM, multiple data bit rates can be transmitted over a single fiber with a total capacity of 100 Gbps or more. This is enough capacity to carry more than one million simultaneous voice conversations. The International Telecommunications Union (ITU) has set the standard wavelength for the lightwaves in a WDM system at 1500 nanometers to 1535 nanometers, which can be divided into 43 channels. (A nanometer is a millionth of a millimeter). Higher-wavelength WDM systems that operate in the 1550 nanometer window are sometimes called *dense wave division multiplexing (DWDM)* systems.

Another important multiplexing technique, time division multiplexing (TDM), does not operate efficiently at speeds higher than 10 Gbps. Unlike WDM, which transmits signals on all channels simultaneously, TDM multiplexes lower bit-rate channels into a single high-speed bit stream. Each channel travels at this higher bit rate for a short duration of time, after which another channel transmits, again at the higher bit-rate. By alternating transmissions between the lower-speed channels, each transmission ultimately is sent at a higher speed. In contrast, using WDM, the different channels are multiplexed together to share a common fiber line simultaneously.

To better understand the differences between TDM and WDM (or DWDM), a tutorial published by the International Engineering Consortium (IEC), Chicago, IL and Lucent Technologies, uses the following analogy: Think of a single fiber as a multi-lane highway. Traditional TDM systems use a single lane of this highway and increase capacity by moving faster in this one lane. Using DWDM is like accessing the unused lanes on the highway—in other words DWDM systems actually increase the number of wavelengths on a fiber strand. This taps a huge amount of unused capacity in the fiber. Another benefit of DWDM, and optical networking in general, is that the "highway" (fiber) is blind to the traffic it carries: "vehicles" can be ATM, SONET and/or IP packets, the tutorial notes. (The tutorial can be found on the Web at *www.webproforum.com/lucent3/*.)

Depending on the kind of fiber used, WDM can increase capacity as much as 40 times. DWDM systems multiplex up to 8, 16, 32 or 40 wavelengths (separate channels). Each signal can be carried at a different rate (OC-3, OC-12, OC-24, etc.) and in a different format (ATM, SONET, etc.) Using a DWDM system, a network operating with SONET signals at OC-48 (2.5 Gbps) and OC-192 (10 Gbps) can achieve total capacities of 40 Gbps. In the future, DWDM systems will carry up to 80 wavelengths of OC-48, a total of 200 Gbps, or 40 wavelengths of OC-192, for a total of 400 Gbps per fiber. This is enough capacity to transmit 90,000 volumes of an encyclopedia in one second over one strand of fiber, according to the IEC tutorial. The cost of deploying WDM systems has come down over recent years, primarily because lasers can now be built for one-tenth their original cost.

WDM See **wavelength-division multiplexing**.

web browser A user interface (usually graphical) to hypertext information on the World Wide Web or on private data networks such as intranets or extranets. Netscape Navigator and Microsoft Internet Explorer are the two most popular browsers.

Wide Area Information Servers (WAIS) A distributed text searching system that uses the protocol standard ANS Z39.50 to search index databases on remote computers. WAIS libraries are most often found on the Internet. WAIS allows users to discover and access information resources on the network without regard to their physical location. WAIS software uses the client-server model.

wide area network (WAN) A network that connects local area networks (LANs) that are geographically distant—usually in different cities or at least 100 kilometers apart. A WAN usually uses dedicated lines leased from a telephone company or public switched telephone network (PSTN) lines. A metropolitan area network (MAN) is often considered a WAN that operates within a city and its suburbs. WANs are used by organizations with offices in remote locations, or by groups of users in different organizations and locations that have similar interests. WANs sometimes combine different types of physical networks by linking, for example, T-1 networks with X.25 networks and/or Integrated Service Digital Network (ISDN) systems.

Wide Area Telephone Service (WATS) A toll service offering for customer dial-type telecommunications between a given customer [user] station and stations within specified geographic rate areas employing a single access line between the customer [user] location and the serving central office. Each access line may be arranged for either outward (OUT-WATS) or inward (IN-WATS) service, or both. Eight hundred number (800) lines are inbound WATS lines.

wideband This term has many meanings depending upon application. *Wideband* is often used to distinguish it from *narrowband,* where both terms are subjectively defined relative to the implied context. Wideband can refer to the property of a circuit that has a bandwidth wider than normal for the type of circuit, frequency of operation, or type of modulation. The most common definition, in telephony, is a circuit that has a bandwidth greater than a 4 kHz voice channel. Also called **broadband.**

wire center See **central office**.

wireless A transmission system that employs radio frequency (RF) through the air as the medium, as opposed to fixed wires. Cellular, PCS, and other mobile systems are all wireless systems.

wireless local loop (WLL) The use of wireless transmission technologies to provide the "last mile" connection between the wired network and the subscriber. In developing countries, using wireless for the local loop is less expensive than wiring each subscriber. Advanced digital wireless technology can also compete with enhanced twisted pair copper wire (*i.e.*, ISDN, DSL) in delivering high bandwidth connections to end users. Also known as **fixed wireless**.

World Wide Web (WWW) The graphical part of the Internet. An international, virtual-network-based information service composed of Internet host computers that provide on-line information in a specific hypertext format. WWW servers provide hypertext metalanguage (HTML) formatted documents using the hypertext transfer protocol (HTTP). Information on the WWW is accessed with a hypertext browser such as Mosaic, Netscape Navigator, Microsoft Internet Explorer, or Lynx.

X Y Z

xDSL Refers to a series of digital subscriber line (DSL) technologies, such as *A*DSL, *H*DSL, *V*DSL, and *RA*DSL. See **digital subscriber line**.

X.-series Recommendations Sets of data telecommunications protocols and interfaces defined by CCITT Recommendations. Some of the more common X.-series Recommendations are:

> **X.25:** A CCITT Recommendation for public packet-switched communications between a network user and the network itself (see below).
>
> **X.75:** A CCITT Recommendation for public packet-switched communications between network hubs.
>
> **X.400:** An addressing scheme for use with e-mail.
>
> **X.500:** An addressing scheme for directory services.

X.25 packet-switched network A connection-oriented packet service for wide-area networks that was dominant for nearly two decades. Newer packet technologies include ATM, frame relay and SMDS. Properly speaking, X.25 is not a kind of network, although the term has come to be used that way. Technically, X.25 is an internationally accepted specification for the interface between data terminal equipment (DTE), such as computer terminals and multiplexers, and data communications equipment (DCE), such as modems and routers. In an X.25 network, perhaps more properly called a packet-switched network (or PSN), DTEs connect to DCEs, which, in turn, connect to a switch (or packet-switching exchange [PSE]), and, finally, to another DTE, such as a computer. X.25 networks can provide either permanent virtual circuits (PVCs), used when data being transmitted is nearly continuous, or switched virtual circuits (SVCs), used when data flow is intermittent. X.25 packets are fixed in length.

Other Books From Aegis Publishing Group

Aegis Publishing specializes in telecommunications books for non-technical end-users. Inquire about wholesale quantity discounts:

Aegis Publishing Group, Ltd.
796 Aquidneck Ave., Newport, RI 02842
800-828-6961; aegis@aegisbooks.com; Web: www.aegisbooks.com

Data Networking Made Easy
The Small Business Guide to Getting Wired for Success, by Karen Patten
$19.95 1-890154-15-6 250 pages

Digital Convergence
How the Merging of Computers, Communications, and Multimedia Is Transforming Our Lives, by Andy Covell
$14.95 1-890154-16-4 240 pages

Strategic Marketing in Telecommunications
How to Win Customers, Eliminate Churn, and Increase Profits in the Telecom Marketplace, by Maureen Rhemann
$39.95 1-890154-17-2 280 pages

Telecom Business Opportunities
The Entrepreneur's Guide to Making Money in the Telecommunications Revolution, by Steve Rosenbush
$24.95 1-890154-04-0 320 pages

Winning Communications Strategies
How Small Businesses Master Cutting-Edge Technology to Stay Competitive, Provide Better Service and Make More Money, by Jeffrey Kagan
$14.95 0-9632790-8-4 219 pages

Telecom Made Easy
Money-Saving, Profit-Building Solutions for Home Businesses, Telecommuters and Small Organizations, by June Langhoff
$19.95 0-9632790-7-6 400 pages

The Telecommuter's Advisor, 2nd edition
Real World Solutions for Remote Workers, by June Langhoff
$14.95 1-890154-10-5 240 pages

900 Know-How
How to Succeed With Your Own 900 Number Business, by Robert Mastin
$19.95 0-9632790-3-3 350 pages

The Business Traveler's Survival Guide

How to Get Work Done While on the Road, by June Langhoff
$9.95 1-890154-03-2 128 pages

Getting the Most From Your Yellow Pages Advertising

Maximum Profits at Minimum Cost, by Barry Maher
$19.95 1-890154-05-9 304 pages

Money-Making 900 Numbers

How Entrepreneurs Use the Telephone to Sell Information
by Carol Morse Ginsburg and Robert Mastin
$19.95 0-9632790-1-7 336 pages

How to Buy the Best Phone System

Getting Maximum Value Without Spending a Fortune, by Sondra Liburd Jordan
$9.95 1-890154-06-7 136 pages

1-800-Courtesy

Connecting With a Winning Telephone Image, by Terry Wildemann
$9.95 1-890154-07-5 144 pages

The Telecommunication Relay Service (TRS) Handbook

Empowering the Hearing and Speech Impaired, by Franklin H. Silverman, Ph.D.
$9.95 1-890154-08-3 128 pages

The Cell Phone Handbook

Everything You Wanted to Know About Wireless Telephony
(But Didn't Know Who or What to Ask), by Penelope Stetz
$14.95 1-890154-12-1 240 pages

Phone Company Services

Working Smarter with the Right Telecom Tools, by June Langhoff
$9.95 1-890154-01-6 102 pages